T0225194

Integrated Business Information Systems

Klaus-Dieter Gronwald

Integrated Business Information Systems

A Holistic View of the Linked Business
Process Chain ERP-SCM-CRM-BI-Big Data

 Springer

Klaus-Dieter Gronwald
Lucerne University of Applied Sciences and Arts
Lucerne School of Business
Lucerne, Switzerland

ISBN 978-3-662-57127-9   ISBN 978-3-662-53291-1   (eBook)
DOI 10.1007/978-3-662-53291-1

Translation of the German language edition "Integrierte Business-Informationssysteme". © Springer-Verlag 2015.

© Springer-Verlag GmbH Germany 2017
Softcover reprint of the hardcover 1st edition 2017

Printed on acid-free paper

This Springer imprint is published by Springer Nature
The registered company is Springer-Verlag GmbH Germany
The registered company address is: Heidelberger Platz 3, 14197 Berlin, Germany

This book is dedicated to Inge, Alexandra, Victoria

Preface

Standardized IT-supported information systems and business processes are influencing each other. They are integrated parts of entrepreneurial thinking and action. Especially the latest developments of big data methodologies and in-memory computing are technology-driven innovations which have a significant impact on corporate structures and the competitive situation of companies. McAfee and Brynjolfsson (2012) call Big Data *The Management Revolution*.

Integrated Business Information Systems mutate more and more into *Intelligent Business Integration Systems*.

Enterprise Resource Planning (ERP), Supply Chain Management (SCM), Customer Relationship Management (CRM), Business Intelligence (BI), and Big Data Analytics (BDA) are business-related tasks and processes, which are supported by standardized software solutions. This requires business-oriented thinking and acting from IT specialists and data scientists.

Timothy Leonard, consultant and data scientist (Eckerson 2012), wrote: *"I rose up through the technical ranks and learned the hard way that you can't be perceived as an IT person. You need to be perceived as a business person who uses technology to solve business problems."*

This quote can be applied directly to computer scientists when they deal with business process-related tasks. It is a good idea to let students experience this directly from the business perspective, for example, as executives of a virtual company in a serious gaming environment. The course simulates the stepwise integration of the linked business process chain ERP-SCM-CRM-BI-Big Data of four competing groups of companies. The course participants become board members with full P&L responsibility for business units of one of four beer brewery groups each from production to retailer.

The story is a combination of facts and fiction. Global and local beer markets are occupied by beer giants. Four investor groups have acquired some of the independent breweries including their entire supply chains from retailer to production (Alpha Beer, Green Beer, Royal Beer, and Wild Horse Beer). Each group has four retail chains distributed all over the country. There is a typical post-merger situation right after the foundation of the four groups with business units having different business processes, product portfolios, rules, tools, and IT infrastructures. With the strategic goals of an ERP implementation (standardizing business processes, standardization of master data, optimization of the IT infrastructure), the post-merger situation will be cleared. The next step is to optimize the supply chains introducing Supply Chain Management (SCM) techniques. With a focus on sales and marketing, Customer Relationship Management (CRM) is implemented initiating the direct competition of the four groups. Real-time Big Data Analytics is the final step for the successful implementation of Integrated Business Information Systems. Role play and gaming phases alternate gradually, starting with the formation of the business units and the analysis of the initial business situation.

The virtual gaming environment *www.kdibis.com* is the web-based business simulation system created specifically for these courses. It is complementary to the book with templates for decisions and presentations including simulation result.

Beer is the ideal product for a supply chain game: The production process is easy to understand; it consists of few main ingredients and has a fixed production cycle of 7 days. It can easily be customized in a variety of products (bottle, flip-top bottle, party keg, 6-pack, . . .) and can easily be distributed. There is a continuous demand with moderate variations, seasonal and random, which guarantees a good dynamic for forecasting and inventory management as essential supply chain management components. The retail groups can also be diversified with different business and distribution models, from beverage markets and retail shops to beer boutiques, each with a specific product portfolio.

The classical *Beer Game* or *Beer Distribution Game*, developed in the 1960s by the Sloan System Dynamics Group (Sterman 1987) to teach scientific principles in management, is used today mainly to experience the *Bullwhip Effect*, a phenomenon in noncommunicative supply chains (Riemer 2012) which has a significant impact on the inventory costs. It is the initial game phase of the *kdibis game*. In contrast to the traditional Beer Game with one company, a simple four-stage supply chain (retailer, distributor, wholesaler, factory), and one product (beer), kdibis has four business groups each with a four-stage supply chain and four retail chains with eleven products each. For each product, for each retailer, a random demand is generated. Additionally, one of four annual consumption distributions, also randomly selected, is superimposed. These distributions correspond to real data from the countries United States, Germany, Austria, and Switzerland as well as the average consumption fluctuations in demand. This is done for each of the four teams. As a result, each company has its own initial data distribution, revenue, and market share.

The structure of the book follows a step-by-step implementation and integration of ERP, SCM, CRM, BI, and Big Data Analytics from a business and project manager's perspective.

The instructor is included into the role play as chairperson discussing the performance of each team in formal board review meetings. Ideal team sizes are between 20 and 30 students with four companies and two students per role. The ideal size is 20, four teams with one person per role. The ideal classroom would be a room with four separate round or squared tables for eight to ten people. Those "learning islands," one for each company, have turned out to be extremely important for an immediate identification of the groups with their company, developing their own group dynamics from day one, while sensing the other groups in the room. It is a noisy experience with a lot of emotions, laughter, and fun.

Online games and role play are optional. The book can be used alone as structured textbook where Part I serves as project guide for the step-by-step introduction of the respective information systems. However, there are no direct exercises for each chapter. Here, too, it is assumed that the course participants present their respective work results and company decisions during the course. Corresponding exercises can be generated from the material in Part III. Here ends the scope of this book.

The story continues with phase five of the implementation of the process chain ERP-SCM-CRM-BI-BIG DATA in the book *Global Communication and Collaboration – Global Sourcing, Global Project Management, Cross-Cultural Competencies* (Gronwald 2017): outsourcing of IT and Business Services. The four beer groups Alpha Beer, Green Beer, Royal Beer, and Wild Horse Beer have grown into global beer giants with a global presence on almost all continents. They are the avatars for Anheuser-Busch InBev, Carlsberg, Heineken, and SABMiller. All of them have outsourced their IT and Business Services to captive centers. Alpha Global IT & Business Services, Green Global IT & Business Services, Royal Global IT & Business Services, and Wild Horse Global IT & Business Services have become profit centers in our simulation. Additionally, they have decided to offshore parts of their services to one or two India-based global service providers (gdigservices and idktech).

Students will become the leadership teams with roles as service cluster heads for ERP, SCM, CRM, and Big Data Analytics (BDA) headed by a Program Manager.

I would like to thank the Lucerne University of Applied Sciences and Arts, which has supported this project with special funding, and has thus committed to new teaching and learning methods. Thank you to all my students at the University of Applied Sciences and Arts Northwestern Switzerland and Lucerne University of Applied Sciences and Arts, who helped me find the right way, and I apologize to all of you, who suffered from some of my experiments gone wrong.

Rotkreuz, Luzern, Switzerland   Klaus-Dieter Gronwald

References

Eckerson W (2012) Secrets of analytical leaders: insights from information insiders. Technics Publications, LLC, Westfield. http://searchbusinessanalytics.techtarget.com/feature/Data-driven-culture-helps-analytics-team-generate-business-value. Accessed 16 Nov 2014

Gronwald K (2017) Global communication and collaboration – global project management, global sourcing, cross-cultural competencies. Springer, Berlin

McAfee A, Brynjolfsson E (2012) Big data: the management revolution. Harvard Business Review. https://hbr.org/2012/10/big-data-the-management-revolution/ar. Accessed 16 Nov 2014

Riemer K (2012) Bullwhip effect. The University of Sydney. http://www.beergame.org/the-game/bullwhip-effect. Accessed 16 Nov 2014

Sterman J (1987) Modeling managerial behaviour: misperception of feedback in a dynamic decisionmaking experiment. Sloan School of Management, Massachusetts Institute of Technology, Cambridge, MA

Contents

List of Figures

List of Tables

Chapter 1
Introduction

Abstract Enterprise Resource Planning (ERP), Supply Chain Management (SCM), Customer Relationship Management (CRM), Business Intelligence (BI) and Big Data Analytics (BDA) are business related tasks and processes, which are supported by standardized software solutions. This requires business oriented thinking and acting from IT specialists and data scientists. It is a good idea to let students experience this directly from the business perspective, for example as executives of a virtual company in a serious gaming environment.

This course simulates the gradual implementation and integration of business information systems and processes as described above of four competing groups of companies over a period of 3–4 fiscal years. As new members of the Management Board, the participants receive the full profit and loss (P & L) responsibility for business areas for each of four beer brewer groups from production to retailer. It is quite common for such projects to be managed by business managers and not by the IT (Killing 2010). This includes the deliberate acceptance of technical incompetence at the beginning (Killing 2010). In so far, the course concept *learning by doing* is quite realistic.

The story is a combination of facts and fiction. Global and local beer markets are occupied by beer giants. Four investor groups have acquired the independent breweries including their entire supply chains (Alpha Beer, Green Beer, Royal Beer, Wild Horse Beer). Each group has four retail chains distributed all over the country. There is a typical post merger situation right after the foundation of the four groups with business units having different business processes, product portfolios, rules, tools and IT infrastructures. With the strategic goals of an ERP implementation (standardizing business processes, standardization of master data, optimization of the IT infrastructure) the post merger situation will be cleared. The next step is to optimize the supply chains introducing Supply Chain Management (SCM) techniques. With a focus on sales and marketing Customer Relationship

© Springer-Verlag GmbH Germany 2017
K.-D. Gronwald, *Integrated Business Information Systems*,
DOI 10.1007/978-3-662-53291-1_1

Management (CRM) is implemented initiating the direct competition of the four groups. Real time Big Data Analytics is the final step for the successful implementation of Integrated Business Information Systems. Roleplay and gaming phases alternate gradually, starting with the formation of the business units and the analysis of the initial business situation.

The virtual gaming environment www.kdibis.com is the web based business simulation system created specifically for these courses. It is complementary to the book with templates for decisions and presentations including simulation result. An integrated webinar system for team presentations and lectures enables the courses to be run as e-learning classes with virtual teams.

The book has three parts.

Part I: Role based business simulation. Introduction of the role based simulation environment kdibis, the story, methods and background information for the online simulation including the registration process and game initiation. It is structured according to the project phases and contains just the basic learning content needed for performing each phase. The use of the online gaming environment is optional. In this case, this part can serve as the master project plan for the implementation of the integration objectives. However, major course objectives (and quite a lot of fun) are lost, when the teams are experiencing failure and success while optimizing their supply chains and in CRM when competing against each other making the right marketing decision.

Part II: Detailed course content and theory for all four phases for building competency. It is structured traditionally according to the topics ERP, SCM, CRM, BI, Big Data Analytics.

Part III: Complementary course material including case studies and company profiles of the virtual kdibis world which are needed to develop an ERP and post-merger strategy.

All other course material like templates for presentations and review meetings are available as downloads from kdibis. This includes complete lecture presentations for each phase, supervisor manuals, additional case studies and a media folder with video clips in English and German.

Reference

Killing P (2010) Nestlé's globe program (A): the early months. Harvard Business Publishing, Boston, MA. IMD 194

Part I
Role Based Business Simulation

Abstract

This part contains the step-by-step implementation of ERP, SCM, CRM, BI and Big Data Analytic as a combined roleplay and virtual business game. This part is to be understood as a master project plan for the implementation of the integration targets in a post merger situation and contains just the basic learning content needed for performing each phase. All details needed for the development of the required competences will be covered in Part II parallel to each project phase.

Part I
Role Based Business Simulation

Abstract

Chapter 2
Preparation and Initiation

Abstract Introduction of the role based virtual simulation environment kdibis, the story, methods and background information for the online simulation including the registration process and game initiation.

2.1 Initial Situation

The local beer market is dominated by two beer giants, which together have a market share of 55% (shrinking). The remaining 45% is shared by a growing number of small craft beer breweries. The beer market is saturated and shrinking globally.

Four groups of investors have acquired a part of the independent breweries plus a wholesaler, a distributor and four different retail groups and founded four new groups of companies with their own brands: *Alpha Beer*, *Green Beer*, *Royal Beer* and *Wild Horse Beer* (Fig. 2.1).

Each group has four retail chains distributed throughout the country, as well as a distributor, a wholesaler and a brewery. The establishment of each of the four groups of companies results in a typical post merger situation with business units that are neither organizationally nor technically harmonizing, including different product portfolios.

The leadership teams have the same organizational structure for all four groups: *CEO, Head Retailer, Head Distributor, Head Wholesaler, Head Factory*. These roles are occupied by the students. These report to the chairperson which is the lecturer. There are three roles of the lecturer:

a) as teacher, responsible for the achievement of the learning objectives,
b) as supervisor for the course administration,
c) as chairperson of each company reviewing and approving the decisions of each team.

© Springer-Verlag GmbH Germany 2017
K.-D. Gronwald, *Integrated Business Information Systems*,
DOI 10.1007/978-3-662-53291-1_2

Fig. 2.1 Company logo

2.2 Registration and Roles

2.2.1 Registration at kdibis.com

Regardless of whether the games are actively played during the course, a registration of the course participants at *www.kdibis.com* is mandatory to access templates for review meetings, lecture notes, presentations, case studies and using the webinar system.

2.2.2 Roles

In any case four groups *Alpha, Green, Royal, Wild Horse* will be generated. Five roles per group are required: *CEO, Head Retailer, Head Distributor, Head Wholesaler, Head Factory* (Fig. 2.2). There can be up to two team members per role. This results in the following scenarios:

a) 5–10 students: one active group, three groups played by the computer
b) 11–15 students: two active groups, two groups played by the computer
c) 16–19 students: three active groups, one team played by the computer
d) 20–25 students: four active groups

For courses with more than 25 students, additional classes can be generated in kdibis. That allows a virtually unlimited number of students per course. The ideal number of students per course is 20.

2.3 Preparation

To log in to the *kdibis Business Game* as integrated part of this text book you must be registered either as a supervisor or a student. The supervisor registration is not required for participation in a course. The student authorization will be carried out by the supervisor.

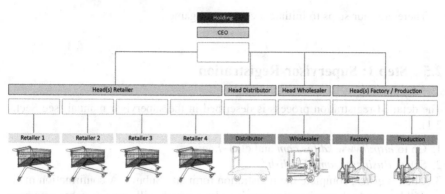

Fig. 2.2 Orgchart

To register as a supervisor this book is mandatory. Vice versa the full utilization of this book requires the game.

The complete registration process is described in a separate online manual as download:

1. Go to *http://www.kdibis.com* and click *Login*.
2. Select *Business Game English*.
3. Login as *Guest* and enter the lobby in the brewery building through the *Brewery* door.
4. In the lobby click the *kdibis logo* right of the elevator door to get to the *visitor center*.
5. Go to the *check-in*, select downloads and download the supervisor manual. There you will find detailed instructions for the registration, for building your own course environment and how to run the Games.

2.4   Game Structure and Organization

Games are organized in *classes*. Each supervisor account can create a virtually unlimited number of classes.

Each class can create a virtually unlimited number of *games*, but only one game can be active at any time. Games can be archived and reloaded if needed.

Each class has a number of *registered students*. Students are organized in *teams* (*Alpha, Green, Royal, Wild Horse*). Each *team member* has a specific *role* (*CEO, Head Retailer, Head Distributor, Head Wholesaler, Head Factory*). The ideal team size is five. There can be more than one member per role, ideally not more than two. All roles have the same privileges except the CEO.

Supervisor are identified by their email address. There can be one supervisor registration per email address only. The same email address can be used as user name to be registered in a virtually unlimited number of classes as student, but only in one role per class.

There are four steps to initiate a course and game.

2.5   Step 1: Supervisor Registration

The detailed registration process is described in the supervisor manual (see Sect. 2.1):

1. Go to *http://www.kdibis.com* and click *Login*.
2. Select *Business Game English*.
3. Click *register*, complete the registration form and Submit. A confirmation mail will be sent to the registration email address which will serve as user name and an automatically generated passcode. This passcode cannot be changed.
4. With clicking the activation link the supervisor registration is completed.
5. Since the free unlimited use of the kdibis gaming site is linked to this text book, an *eligibility check* is required the first time a supervisor is logging in. A random set of four images from the book will have to be identified with their correct figure number in the right sequence. If one of the answers is not correct, a new set of four images will be generated.

2.6   Step 2: Creating Classes

The detailed registration process is described in the supervisor manual (see Sect. 2.1):

1. Go to *http://www.kdibis.com* and click *Login*.
2. Select Business Game English.
3. Login as supervisor and enter the lobby in the brewery building through the *Brewery* door. That selects English as language.
4. In the lobby click the *Alpha Beer* logo on the left side of the elevator. Click *Access Control*. When your login was successful you should be identified as *Chairperson* and the palm scanner left should be green. Click the green button, go to the offices, and click the *Guest Beer* icon on the screen.
5. In the supervisor menu click administrator and select *class admin*, then *create class*.
6. Enter a Class Title and Submit. When successful, a registration mail was sent to the supervisor with an activation link.
7. Once done, go back to the supervisor menu and select the just generated class as active class. If there are more than one classes, the actual active class will be active during the entire session.

   See the supervisor manual for more class admin options.

2.7   Step 3: Creating Games

The detailed game creation process is described in the supervisor manual (see Sect. 2.1):

1. In the supervisor menu click *administrator* and select *game admin*, then *new game*.
2. Select the active class from the menu. If there is no class, go to Step 2 (Sect. 2.6).
3. Select one of the four game types *SCM1, SCM2, CRM1, or CRM2*. Enter *Institution* and a *Game Title* and *Create Game*. Games will be created for all four teams.
4. See the supervisor manual for more game admin options.

2.8   Step 4: Register Students

The detailed registration process is described in the supervisor manual (see Sect. 2.1):

1. In the supervisor menu click *administrator* and select *student admin*, then *register student*.
2. Enter the student email.
3. Once the students have completed the activation with entering first name and name, the supervisor will have to assign them to a game and a role.
4. In the *student admin* menu select *student role*.
5. Select the team (*Alpha Beer, Green Beer, Royal Beer, Wild Horse Beer*).
6. Select a role (*CEO, Retailer, Distributor, Wholesaler, Factory*) and *Submit*.

2.9   Student Login

Once students have registered for an active game and assigned to a team and role, they can login with their credentials:

1. http://www.kdibis.com $\Rightarrow$ login $\Rightarrow$ Business Game English.
2. Login with email address and password.
3. Enter the lobby in the brewery building through the *Brewery* door. This will activate English as game language.
4. In the lobby click the logo of your company (*Alpha, Green, Royal, Wild Horse*) on the left side of the elevator. Click *Access Control*.
5. When your login was successful you should be identified with your role and the palm scanner left should be green.
6. Click the green button, go to the office, and click the icon of your company on the screen. The student cockpit gives access to a set of modules which will be

activated depending on the game type (*SCM1*, *SCM2*, *CRM1*, *CRM2*). Check the supervisor manual for details (see Sect. 2.1).

2.10 Kdibis Webinar System

Students of an active class can use the kdibis webinar system for presenting results to team members and share results. Lecturers can run lectures with the online course material or using their own presentations and download student presentations for review and grading. This enables the course to be run as e-learning class remotely with virtual teams.

1. Login as student or supervisor to your class.
2. In the lobby go to the visitor center and the *check-in*.
3. To upload and present a presentation, go to the *Meeting Room* of your team.
4. To view a presentation, visit the *Meeting Room* of the corresponding team.

Check the supervisor manual for details (see Sect. 2.1).

Chapter 3
Development and Implementation of an ERP Strategy

Abstract There is a typical post merger situation right after the foundation of the four groups with business units having different business processes, product portfolios, rules, tools and IT infrastructures. With the strategic objectives of an ERP implementation (standardization of business processes, standardization of master data, optimization of the IT infrastructure), this situation is adjusted accordingly for the four groups. The analysis of the business situation based on the results of the previous fiscal year and a product portfolio analysis are the basis for a first review meeting where the teams present their strategies.

3.1 Situation Analysis

The new management teams of the four groups are all facing the same problem of the integration of independent, individual companies, starting with each retail group, through distributors, wholesalers, breweries to production; eight interdependent, previously independently acting business areas with full P & L responsibility and with their own past. These are documented in detail in Part III for each part of the company and as downloads in the *Document* folder. They are subdivided into:

- Market—distribution structure—customers
- IT infrastructure
- Product portfolios
- Results of the previous fiscal year

The following problem areas were identified at a first glance:

- Current business situation
- M&A IT integration strategy
- Business process analysis, especially the *order-to-cash*
- Product portfolio analysis

© Springer-Verlag GmbH Germany 2017
K.-D. Gronwald, *Integrated Business Information Systems*,
DOI 10.1007/978-3-662-53291-1_3

3.2 ERP Strategy

Sales structures, business processes and IT infrastructures are linked via ERP and can thus be standardized and implemented using a suitable ERP strategy and appropriate software systems. The operational view of an ERP system describes it *as a system that supports all business processes of a company. It contains modules for procurement, production, sales, asset management, finance and accounting etc., integrated via one common database* (Springer Gabler Verlag [1]).

The inclusion of business objectives in the considerations leads to a more strategic approach for the implementation goals of an ERP system. *The standardization of business processes beyond organizational boundaries can have significant synergy effects. Organizations can implement best practices in the system, and the ERP system is perceived as a business tool rather than an IT tool* (Desai and Srivastava 2013).

During the last 10–15 years, global companies have successfully implemented ERP from a business perspective to achieve best practices and achieve global synergies. One of the largest projects of its kind is Nestlé GLOBE (Nestlé 2014). The project began in 2000. Nestlé's CEO, Peter Brabeck, then defined the GLOBE objectives as follows: *„I want this to be very clear. With GLOBE, we will create common business processes, standardized data, and a common IT infrastructure—but do not think this is an IT initiative. We are going to fundamentally change the way we run this company."* (Killing 2010).

With this Brabeck formulated the three fundamental business objectives of an ERP implementation:

- Generating a standardized business process architecture
- Standardization of internal and external master data
- Standardization of the IT infrastructure

These three objectives are to be formulated here by the teams and transferred to their company situation (see Part II, Chap. 10 for details).

3.3 M&A IT Integration

The elimination of inhomogeneous IT environments caused by a merger is part of the ERP strategy (see Part II, Chap. 10 for details). The current IT situation for all divisions can be found in Part III, Chaps. 15–18. This is the basis for the decisions to be made here.

In principle, four scenarios can be distinguished:

S1—Coexistence/Symbiosis
Retain both IT systems. Build portal above current systems to aggregate information. The focus is on business process standardization and master data

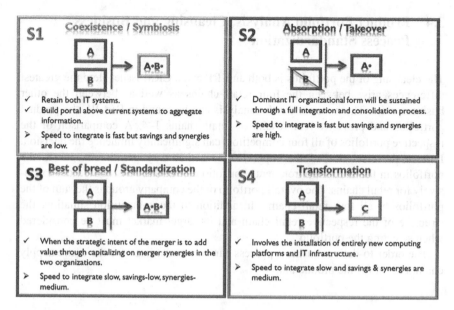

Fig. 3.1   M&A IT integration

consolidation. This is required for all existing systems separately and needs additional effort. Speed to integrate is fast but savings and synergies are low (Fig. 3.1).

S2—Absorption/Takeover
The dominant IT organizational form will be sustained through a full integration and consolidation process. This is used to develop an ERP template, which is implemented in all business units. All three primary strategic ERP objectives will be achieved. However, this requires a great effort of *business process reengineering* in all business areas in which the new system is implemented, combined with appropriate training and *organizational change management*. Speed to integrate is fast but savings and synergies are low (Fig. 3.1).

S3—Best of Breed/Standardization
It realizes the strategic intent to add value of a merger through capitalizing synergies in all organizational units. The result is also an ERP template, which, however, represents the synergy of the best practices of all business units and thus requires *business process reengineering* and *organizational change management* at all levels and in all organizations. Speed to integrate is slow, savings are low with medium synergies (Fig. 3.1).

S4—Transformation
The implementation of a completely new IT platform and infrastructure while replacing existing systems is the most comprehensive and elaborate of the four strategies. The result is also an ERP template. This strategy makes sense when existing systems are outdated or they cannot meet the new requirements. Speed to integrate is slow and savings & synergies are medium (Fig. 3.1).

3.4   Product Portfolio Analysis, Cleansing and Business Process Standardization

The cleansing of the portfolios is both an ERP and a CRM issue. Here the greatest differences arise between the four retail chains as well as between the other business units. The material for the analysis can be found in Part III, the product portfolios and previous year results of retail chains 1–4. A comparison of the respective portfolios of all four competitors can significantly influence the decision. It is at the discretion of the supervisors whether they permit changes to the portfolios in the kdibis environment. Portfolio consolidation is carried out exclusively for retail chains. The overall portfolio of the company group is the sum of the portfolios of all four retail chains. In addition to the competition situation, the structure of the respective retail chain and its target market must be considered when selecting the portfolio.

The order-to-deliver business process can be standardized for the entire supply chain, except for the production.

3.4.1   Order-to-Deliver Process Retail Chain1—Distributor

The retail chain1 supplies its own stores and large retail chains with the entire portfolio, generally except kegs. In addition to beverages, they are selling food and snacks in their own shops (Fig. 3.2).

3.4.2   Order-to-Deliver Process Retail Chain2—Distributor

The retail chain2 supplies own beverage markets with the entire portfolio except single bottles. They sell kegs, 6-packs and cases to private customers and restaurants (Fig. 3.2).

3.4.3   Order to Deliver Process Retail Chain3—Distributor

The retail chain3 operates own beer boutiques in malls and exclusive shopping streets with their own brand. They sell to private customers only. Cases and kegs will be delivered to private or business events for select customers only (Fig. 3.2).

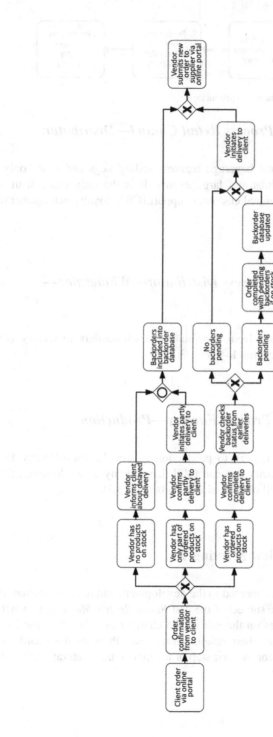

Fig. 3.2 Order to deliver process

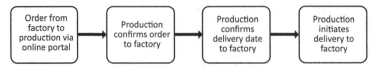

Fig. 3.3 Order to deliver process factory—production

3.4.4   Order to Deliver Process Retail Chain4—Distributor

The retail chain4 operates own beverage markets selling kegs and cases only. Customers are restaurants, clubs and large events. It is the only chain with a dedicated sales team and individual customer support (CRM loyalty management) (Fig. 3.2).

3.4.5   Order-to-Deliver Process Distributor—Wholesaler— Factory

The order-to-delivery processes from distributors to wholesalers to factory are identical to the ones of retail chains 1–4 (Fig. 3.2).

3.4.6   Order-to-Deliver Process Factory—Production

The order-to-deliver process factory to production is straightforward, since the production has unlimited capacity and can fulfil demand any time. However, the delivery delay is identical to all other business units (Fig. 3.3).

3.5   Task—Board Review Meeting 1

The identified problem areas connected to the development and implementation of an ERP strategy (Sect. 3.1) are subject of the first *Board Review Meeting*, in which the supervisor or lecturer takes on the role of the chairperson of each of the four companies. The teams present their results and submit them to the board for approval. These are the basis for the first step of a supply chain integration.

3.5.1 Agenda

1. Analysis of the current business situation based on the consolidated results of the previous year.
2. Proposal for an M&A IT integration strategy including reasoning, action and high level project and implementation plan.
3. Analysis of the order-to-deliver process, SWOT analysis and implementation plan.
4. ERP—business process analysis, master data consolidation, template, localization, organizational readiness.
5. Summary and conclusions.
6. Discussion.

Registered users (faculty and students) can download templates for each team from kdibis.com (see supervisor manual).

References

Desai S, Srivastava A (2013) ERP to E^2RP a case study approach. PHI Learning Private Limited, Delhi

Killing P (2010) Nestlé's globe program (A): the early months. Harvard Business Publishing, IMD 194

Nestlé (2014) GLOBE Center Europe. http://www.nestle.de/karriere/arbeiten-nestle/nestle-international/globe-center-europe. Accessed 16 Nov 2014

Springer Gabler Verlag [1] (Herausgeber) Gabler Wirtschaftslexikon, Stichwort: ERP, online im Internet. http://wirtschaftslexikon.gabler.de/Archiv/3225/erp-v14.html. Accessed 16 Nov 2014

Chapter 4
Game Round 1: Supply Chain Not Optimized—The Beer Game

Abstract In the game round 1 students experience the effects of a non-communicative supply chain with isolated business units which reflects the situation in the first fiscal year without the planned but not yet realized ERP integration strategy. This follows mainly the rules and principles of the classical Beer Game. The results of this round will become the starting point for a supply chain integration and optimization strategy.

4.1 Initial Situation

The decided ERP integration strategy has not yet been implemented. It has little influence on the problems of supply chain management. One fiscal year will be played in the current non-transparent and inconsistent supply chain environment experiencing the problems of a non-optimized supply chain. Each team plays separately. Competition takes place only indirectly by comparison of the final results of this round. Typical results for market shares (Fig. 4.1) and inventory costs (Fig. 4.2) at the end of round 1:

4.2 SCM Game 1: Preparation

The necessary steps for the initiation of the kdibis gaming environment have been described in Chap. 2 and can be found in the supervisor manual. The following preparation steps are needed:

a) Registration
   The registration process for faculty and student has been finished already.
b) Generate games
   Supervisor generate a new game as described in the supervisor manual.
   Select game type *SCM1—not optimized*.

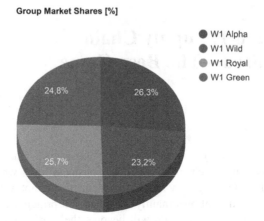

Fig. 4.1 Group market shares

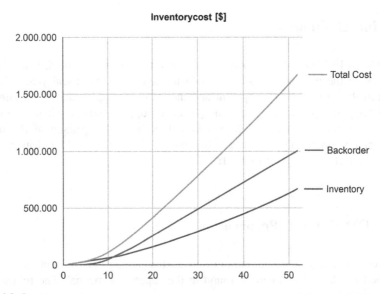

Fig. 4.2 Inventory cost

4.3   SCM Game 1: Execution

Supervisor are starting the game for each team individually (see supervisor manual).

4.3.1 *Fiscal Year, Order Cycle, Delivery Delay*

The *fiscal year* has 364 day. The *order cycle* is fixed to 7 days or 1 week. That results in 52 game rounds corresponding to 52 weeks. The *delivery delay* is 2 weeks.

Cycle 1:

- Orders sent to suppliers (Retailer to Distributor, Distributor to Wholesaler, Wholesaler to Factory, Factory to Production) (Fig. 4.3).
- Order received at suppliers, calculation of delivery depending on actual inventory and delivery to customer initiated (Production to Factory, Factory to Wholesaler, Wholesaler to Distributor, Distributor to Retailer and Retailer to Customer) (Fig. 4.4).

Cycle 2:

- Delivery to customers (Fig. 4.5).

Cycle 1 and 2 are overlapping, resulting in a factual delivery cycle of 7 days in steady state. The game ends after 52 cycles (52 weeks).

Retailer		Distributor		Wholesaler		Factory	
Green Retailer		**Green Distributor**		**Green Wholesaler**		**Green Factory**	
Week	**1**	**Week**	**1**	**Week**	**1**	**Week**	**1**
Order In:	**2'173 hl**	**Order In:**	**2'369 hl**	**Order In:**	**2'217 hl**	**Order In:**	**2'282 hl**
Delivery In:	2'173 hl	Delivery In:	2'369 hl	Delivery In:	2'217 hl	Delivery In:	2'282 hl
Inventory:	2'776 hl	Inventory:	3'316 hl	Inventory:	3'236 hl	Inventory:	2'578 hl
Delivery out:	2'173 hl	Delivery out:	2'369 hl	Delivery out:	2'217 hl	Delivery out:	2'282 hl
Order out:	**2'000 hl**	**Order out:**	**0 hl**	**Order out:**	**0 hl**	**Order out:**	**0 hl**
Backorder:	0 hl	Backorder:	0 hl	Backorder:	0 hl	Backorder:	0 hl
Inventory cost:	$8'883	Inventory cost:	$8'158	Inventory cost:	$6'117	Inventory cost:	$3'765
Backorder cost:	$0	Backorder cost:	$0	Backorder cost:	$0	Backorder cost:	$0
I&B Cost:	**$8'883**	**I&B Cost:**	**$8'158**	**I&B Cost:**	**$6'117**	**I&B Cost:**	**$3'765**

Fig. 4.3 Game cycle 1—order out

Retailer		Distributor		Wholesaler		Factory	
Green Retailer		**Green Distributor**		**Green Wholesaler**		**Green Factory**	
Week	**2**	**Week**	**2**	**Week**	**2**	**Week**	**2**
Order In:	**2'504 hl**	**Order In:**	2'000 hl	**Order In:**	**0 hl**	**Order In:**	**0 hl**
Delivery In:	2'369 hl	Delivery In:	2'217 hl	Delivery In:	2'282 hl	Delivery In:	1'999 hl
Inventory:	2'641 hl	Inventory:	3'533 hl	Inventory:	5'518 hl	Inventory:	4'577 hl
Delivery out:	2'504 hl	Delivery out:	2'000 hl	Delivery out:	0 hl	Delivery out:	0 hl
Order out:	**0 hl**	**Order out:**	**0 hl**	**Order out:**	**0 hl**	**Order out:**	**0 hl**
Backorder:	0 hl	Backorder:	0 hl	Backorder:	0 hl	Backorder:	0 hl
Inventory cost:	$10'203	Inventory cost:	$9'925	Inventory cost:	$8'876	Inventory cost:	$6'053
Backorder cost:	$0	Backorder cost:	$0	Backorder cost:	$0	Backorder cost:	$0
I&B Cost:	**$10'203**	**I&B Cost:**	**$9'925**	**I&B Cost:**	**$8'876**	**I&B Cost:**	**$6'053**

Fig. 4.4 Game cycle 1—order in–delivery out

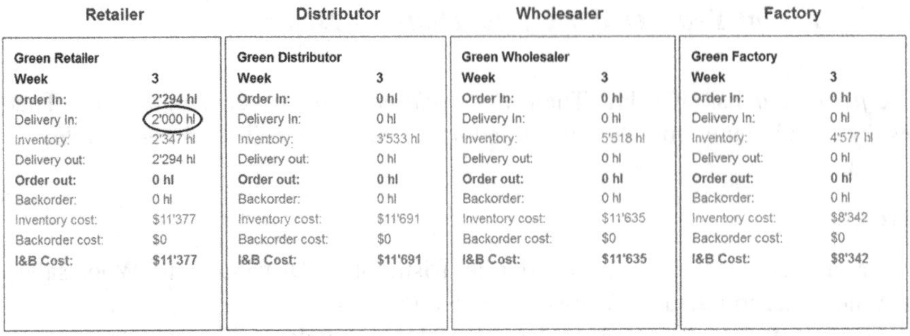

Retailer		Distributor		Wholesaler		Factory	
Green Retailer		**Green Distributor**		**Green Wholesaler**		**Green Factory**	
Week	**3**	**Week**	**3**	**Week**	**3**	**Week**	**3**
Order In:	2'294 hl	Order In:	0 hl	Order In:	0 hl	Order In:	0 hl
Delivery In:	2'000 hl	Delivery In:	0 hl	Delivery In:	0 hl	Delivery In:	0 hl
Inventory:	2'347 hl	Inventory:	3'533 hl	Inventory:	5'518 hl	Inventory:	4'577 hl
Delivery out:	2'294 hl	Delivery out:	0 hl	Delivery out:	0 hl	Delivery out:	0 hl
Order out:	0 hl	Order out:	0 hl	Order out:	0 hl	Order out:	0 hl
Backorder:	0 hl	Backorder:	0 hl	Backorder:	0 hl	Backorder:	0 hl
Inventory cost:	$11'377	Inventory cost:	$11'691	Inventory cost:	$11'635	Inventory cost:	$8'342
Backorder cost:	$0	Backorder cost:	$0	Backorder cost:	$0	Backorder cost:	$0
I&B Cost:	$11'377	I&B Cost:	$11'691	I&B Cost:	$11'635	I&B Cost:	$8'342

Fig. 4.5 Game cycle 2—delivery out

4.3.2   Game Rules

This game round will be played according to *Beer Game rules*. That means that each team member only sees its own business unit (Retailer, Distributor, Wholesaler, Brewery) except the CEO, who supervises the actions of his/her team members and triggers the next week, once all orders have been submitted. There is no forecasting, no inventory management and no communication between team members (Fig. 4.6).

The task is to minimize the inventory and thus the inventory costs, while at the same time maintaining the readiness to deliver.

Orders are fully executed if there is sufficient inventory as the sum of the incoming deliveries (*Delivery In*) and the existing inventory (*Inventory*). If the inventory is less than the purchase order, a partial delivery is made in the amount available. The difference between purchase order and delivery quantity is recorded as a *Backorder*. This will be delivered as soon as there is enough material available. Backorders will accumulate and be reduced or increased according to availability.

Inventory costs are 25% of the selling price per hectoliter. They are cumulated over the fiscal year, as are the backorder costs, which are 50% of the selling price per hectoliter. Since, starting with the production, each profit center has a 30% margin on their respective purchase prices, this means that both inventory costs and backorder costs are different for each area and grow from production to retail.

The only active task the team members can perform is to send orders to their suppliers.

Orders to the retail chains are automatically generated and generated for each individual product in the portfolio. Alpha Beer (Fig. 4.7), Green Beer (Fig. 4.8), Royal Beer (Fig. 4.9), Wild Horse Beer (Fig. 4.10). Each group has the same portfolio totally, but product sales distribution is unique for each group. This will be generated randomly, so each game has a slightly different initial distribution of products in each of the four retail chains.

Fig. 4.6 Orders and deliveries

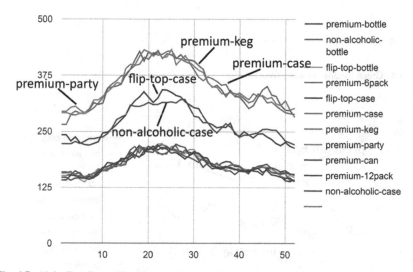

Fig. 4.7 Alpha Retailer product forecast

They will automatically be cumulated to one bulk order per retail chain for Alpha Beer (Fig. 4.11), Green Beer (Fig. 4.12), Royal Beer (Fig. 4.13), Wild Horse Beer (Fig. 4.14).

The orders for the four retail chains are sent to the distributor as one bulk order and from there to the wholesaler, the brewery and the production. It is easy to see that the periodic demand is relatively stable and only slightly fluctuates but is superimposed with a seasonal demand distribution which is significantly larger.

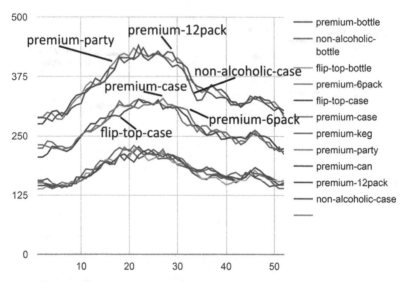

Fig. 4.8 Green Retailer product forecast

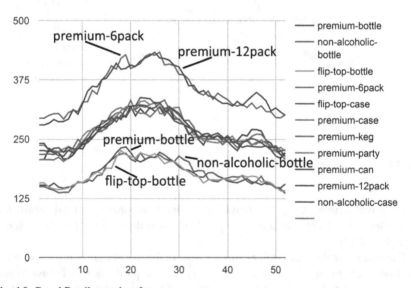

Fig. 4.9 Royal Retailer product forecast

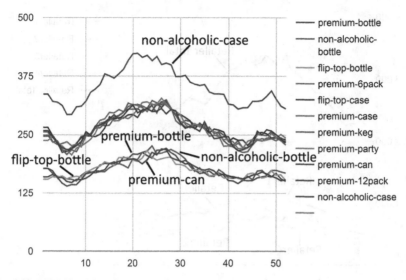

Fig. 4.10   Wild Horse Retailer product forecast

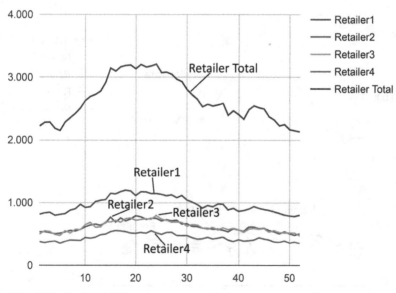

Fig. 4.11   Alpha Retailer 1–4 forecast

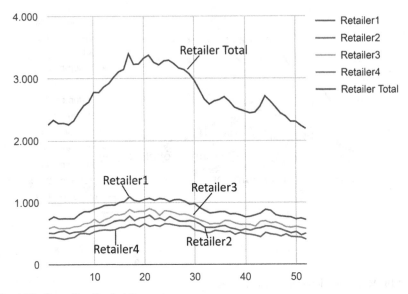

Fig. 4.12 Green Retailer 1–4 forecast

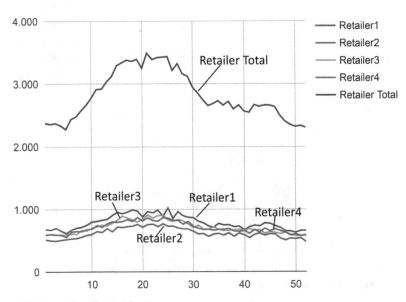

Fig. 4.13 Royal Retailer 1–4 forecast

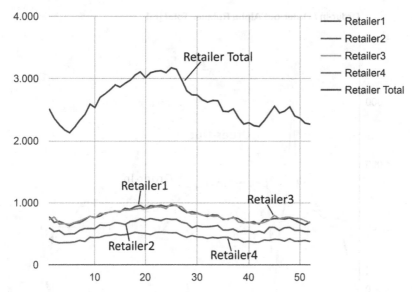

Fig. 4.14 Wild Horse Retailer 1–4 forecast

4.3.3   Results Game Round 1

Typical results from this round clearly show the bullwhip effect in the orders with increasing oscillations from stage to stage and thus the analogous fluctuations in the inventory with the associated costs (Figs. 4.15, 4.16, 4.17 and 4.18).

4.3.4   The Role of the CEOs

Only the CEOs have complete visibility of the supply chains in this round (Fig. 4.19).

They submit the orders once all divisions have placed their orders. Their main task is to observe the order behavior of their team members. However, they are not allowed to communicate with their team members at this stage, but merely observe it. The subsequent team analysis has an impact on the supply chain management strategy.

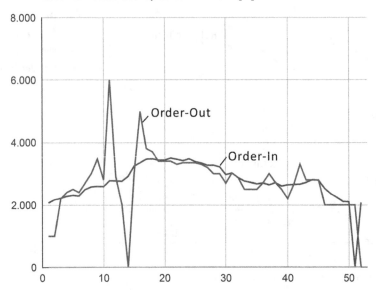

Fig. 4.15 Order-in/order-out Alpha Retailer without SCM

Fig. 4.16 Order-in/order-out Alpha Distributor without SCM

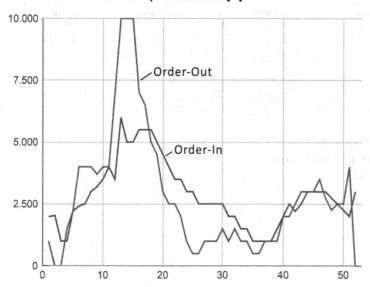

Fig. 4.17   Order-in/order-out Alpha Wholesaler without SCM

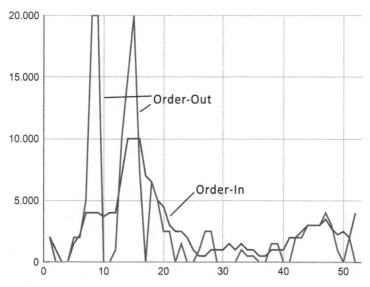

Fig. 4.18   Order-in/order-out Alpha Factory without SCM

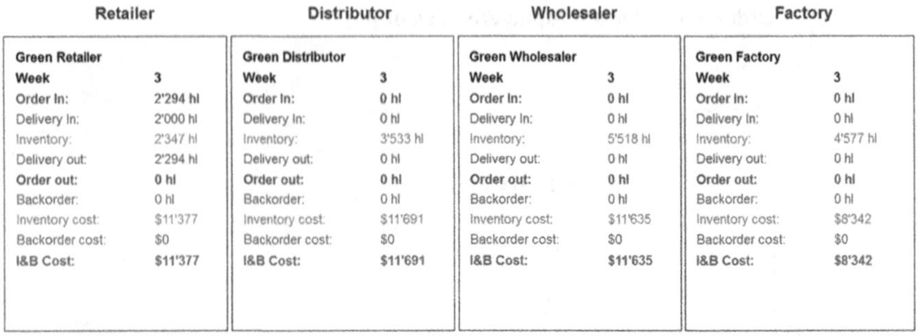

Retailer	Distributor	Wholesaler	Factory

Green Retailer		**Green Distributor**		**Green Wholesaler**		**Green Factory**	
Week	3	Week	3	Week	3	Week	3
Order In:	2'294 hl	Order In:	0 hl	Order In:	0 hl	Order In:	0 hl
Delivery In:	2'000 hl	Delivery In:	0 hl	Delivery In:	0 hl	Delivery In:	0 hl
Inventory:	2'347 hl	Inventory:	3'533 hl	Inventory:	5'518 hl	Inventory:	4'577 hl
Delivery out:	2'294 hl	Delivery out:	0 hl	Delivery out:	0 hl	Delivery out:	0 hl
Order out:	0 hl	Order out:	0 hl	Order out:	0 hl	Order out:	0 hl
Backorder:	0 hl	Backorder:	0 hl	Backorder:	0 hl	Backorder:	0 hl
Inventory cost:	$11'377	Inventory cost:	$11'691	Inventory cost:	$11'635	Inventory cost:	$8'342
Backorder cost:	$0	Backorder cost:	$0	Backorder cost:	$0	Backorder cost:	$0
I&B Cost:	$11'377	I&B Cost:	$11'691	I&B Cost:	$11'635	I&B Cost:	$8'342

Fig. 4.19 CEO SCM cockpit

Chapter 5
Development and Implementation of a SCM Strategy

Abstract Based on the results obtained in game round 1, the theoretical foundations of supply chain management methods are developed and implemented with a corresponding optimization strategy. Supply chain management will be defined and the bullwhip effect will be discussed in detail, including the measures for its prevention. *Demand forecasting* and *inventory management* will be introduced as primary supply chain management methods. The detailed theories and learning content of the presented methods will be discussed in Part II, Chap. 11, mainly in parallel self-study. At a second review meeting, the teams present their results and interpret their behavior from round 1 from which they derive solutions for an optimal communicative supply chain. They must choose both a *forecasting method* and an *inventory management process* that will be implemented for the next game round.

5.1 Supply Chain Management Definition

Supply Chain Management (SCM) is the control of material, information, and financial flows within a supply chain from the raw material supplier through the manufacturer, the intermediate trade to the end customer.

Supply chain management systems synchronize the order-to-cash process, i.e. information streams (orders) with goods and services (deliveries) and cash flows (invoices/payments).

The goal of an efficient supply chain management system is to minimize inventories while ensuring delivery.

5.2 The Bullwhip Effect

Generally, the bullwhip effect can be described as the *oscillation of the demand, which grows with increasing distance from the end customer* (Beer 2014).

This phenomenon of the seemingly irrational ordering behavior of partners within a supply chain, which we could observe in the game round 1, has long been known, but has been systematically investigated only since the end of the 1990s. There is a difference between investigations dealing with the causes and the discussion of solution models. Most authors consider simple linear supply chains to study the causes and solution models for the bullwhip effect (Beer 2014).

Sucky (2009) uses for his analysis a structure of a linear supply chain analogous to our system, which has only several retailers at the retail stage to investigate the effects of risk distribution on the bullwhip effect in a delivery network. Even if our supply chain divides into four parallel strands of retail groups, the ordering and delivery behavior in our system is a linear chain. With the focus on the significant automation of the supply chain with the goal of minimizing inventory costs while maintaining delivery, we will restrict the detailed treatment of the bullwhip effect accordingly to simple linear supply chains.

Similarly, the product portfolios do not play a role in the SCM investigations since they are essentially only other distribution forms for the same product (beer). At this stage, the distinction between premium and non-alcoholic beer is also neglected. However, the products will play a dominant role in the CRM segment.

In many cases, scientific investigations or their authors tend to polarize opinions about the cause and effect of their investigated phenomena, including the bullwhip effect. The isolation of the business segments from the business processes of the entire supply chain leads to different perceptions of the participants, which they only can take from the incoming orders and deliveries, especially if the delivery quantity differs from the order quantity. In the case of supplier bottlenecks, it is lower and leads automatically to backorders (incorrect quantities). These can then lead surprisingly for the ordering business unit to a large subsequent delivery as soon as the supplier can deliver again. Some authors describe the behavior of the participants as *irrational*, especially in the case of supply bottlenecks (Chen et al. 2000, p.270; Holland and Sodhi 2004, p.252; Cronson and Donohue 2006, p.324) others as *rational* behavior (Lee et al. 1997, p.95; Metters 1997, p.9s).

As usual, the truth lies obviously in between. The simulations described here were based on experiences with more than 200 students. Frequently it was the attempt to react rationally with irrational results, especially at the beginning. Especially in the later course of the game, frustration increasingly came into play. The role of CEOs, who have an overview of the order behavior of their teams, is mainly the observation and analysis of the (seemingly) irrational behavior of their team members. This has impact on the development of a supply chain strategy, which should avoid such phenomena.

Of the variant reasons for the occurrence of the bullwhip effect, here only the following parameters are taken into consideration: the delays in delivery, the perceptions, behaviors and actions of the team members, and the non-transparent, isolated, uncoordinated supply chain.

5.3 Demand Forecasting

Demand forecasting and *inventory management* are two key components for optimizing and automating supply chain management. Demand forecasting is primarily used to suppress the bullwhip effect (Chen et al. 2000). Inventory Management is used to optimize inventory while minimizing inventory costs. Order quantities, order cycle, minimum and maximum inventory levels are determined from this.

Forecasting is a significant element of demand management. It is the basis for the planning and the basis for management decisions. The future planning is a stochastic process and therefore an exact prediction is not possible. The goal is therefore to develop techniques that mimic the gap between the current demand and the forecast. Forecasting is therefore an iterative, repetitive process, which compares the prediction with the reality and adapts accordingly.

Demand Forecasting is the link between Demand Generation (CRM: Sales, Marketing) and Demand Fulfillment (SCM: Manufacturing, Logistics) and is therefore a prerequisite for successful supply chain integration with a periodicity of between 1 day and 1 month (depending on industry).

5.3.1 Qualitative Forecasting Methods

Qualitative forecasting methods are based on intuition and experience of the forecaster. They are mainly used for long-term projections, such as the introduction of new products, if current data is not relevant or does not exist. Qualitative methods are for example

- Opinion of an executive committee
- Delphi methods
- Salesforce experience
- Customer surveys

5.3.2 Quantitative Forecasting Methods

Can be separated into two dominating classes (DecisionCraft 2010) *cause and effect methods* and *time-series methods*.

Cause and effect methods (also referred to as causal methods) are used when one or more external factors (independent variables) are related to the demand and thus influence the forecast, such as income development, economic forecasts, etc. (econometric factors). They sometimes require elaborate statistical methods (regression, econometric models, neural networks). These procedures are covered in the section *Big Data Demand Management*.

Time-series methods assume that the future is a continuation of the past. Historical data are used for predicting the future without considering external factors. In this chapter, we deal exclusively with *time-series methods*. In detail, they are

- Naive forecasting
- Simple moving average forecasting
- Weighted moving average forecasting
- Exponentially smoothing forecasting
- Linear trend forecasting

Common to all these processes is their smoothing effect on fluctuations in demand.

5.4   Inventory Management

Inventory costs add to an average of 30–35% of the material value of a product or 6–15% of the revenue depending on the industry (Dietl 2012).

Inventory cost causes are according to Waser (2010)

- **Bound capital**: Purchasing and storage must be pre-financed. For the bound capital interest on current assets must be paid.
- **Infrastructure and handling costs**: Storage infrastructure (including storage and removal) of goods must be made available. Creation, maintenance and depreciation costs.
- **Material management costs**: Material stocks must be regularly checked and re-ordered as required.
- **Depreciation due to aging**: Material in stock loses value. These value adjustments can be significant (date of expiry or innovation).
- **Backorder costs**: Inventories that are too low lead to backorder costs (out-of-stock) in the form of expensive express deliveries or lost sales.
- **Concealed defect costs**: Inventory balances the production processes and thereby obscures potential weak points in production and logistics (kanban, just-in-time).

5.4.1   Inventory Management Models

According to Dietl (2012) we can distinguish between *single-period models* and *multi-period models*, which can have both *deterministic* and *stochastic* demand structures:

- **Single-period models with deterministic demand** are contractually regulated sales volumes within one period (newspaper subscription).

- **Single-period models with stochastic demand** have an unpredictable demand within one period (newspaper sales in shops).
- **Multi-period models with deterministic demand** are contractual defined sales volumes over several periods (supplier contract).
- **Multi-period models with stochastic demand** have an unpredictable demand over several periods (beer sales in retail shops).

Additionally, we distinguish between

- **Fixed-order quantity model** with fixed (optimal) order quantity depending on the target inventory. The order date is variable when the minimal defined inventory is reached.
- **Fixed-time period model** with variable order quantity, depending on the current consumption. The order time is periodic, at a fixed time interval.

This course uses a *fixed-time period model* with *stochastic demand*.

5.4.2 Economic Order Quantity Models (EOQ)

The method, originally developed by Harris (1913) and Andler (1929) as one formula for determining the *optimal production quantity (lot size),* is now mainly used as *Economic Order Quantity Model (EOQ)* to determine the *optimal order quantity* while *minimizing inventory costs.* There are a number of extensions to the basis model. In this course, three models are used, whereby the course participants decide for one of the models based on their results from the first fiscal year (game round 1) and thus must determine the optimal parameters for the supply chain management automation.

The following inventory costs will be taken into consideration here:

- **Carrying costs**—direct costs caused by material storage.
- **Ordering costs**—costs to replenish the inventory.
- **Shortage costs (backorder costs)**—costs if demand cannot be fulfilled due to lacking inventory.

5.4.3 The Basic EOQ Model

The Basic EOQ Model is a single formula for determining the optimum order size, which minimizes the sum of inventory costs and order costs (Fig. 5.1).

Assumptions:

- Demand is known and relatively constant over time.
- No shortages allowed.
- Constant lead time.
- The order quantity is delivered completely.

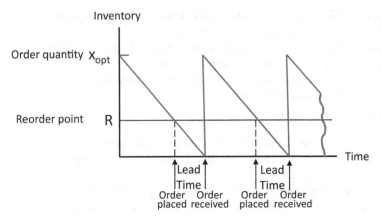

Fig. 5.1 EOQ basic model

5.4.4   Non-instantaneous Receipt Model

The order quantity is delivered partially, distributed over the period between two ordering cycles. This results in different values for inventory costs and inventory (Fig. 5.2).

Assumptions:

- Demand is known and relatively constant over time.
- No shortages allowed.
- Constant lead time.
- The order quantity is delivered partially, distributed over the period between ordering cycles.

  Special parameters for this model:
  p = daily rate where the inventory is replenished.
  d = daily rate where the inventory is depleted.

5.4.5   Shortages Model

The basic EOQ model does not allow shortages. The shortages model does allow this explicitly. However, it is assumed that the total demand will be delivered including the shortages as backorder (Fig. 5.3).

Assumptions:

- Demand is known and relative constant over time.
- No shortages allowed.
- Constant lead time.

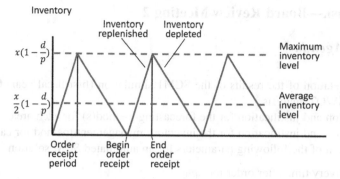

Fig. 5.2 Non-instantaneous receipt model

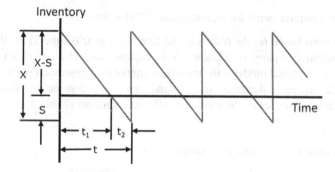

Fig. 5.3 EOQ shortages model

- The order quantity is delivered completely.

Since backordered demand or shortages (S) are balanced as soon as the inventory is filled, the maximum inventory never reaches X, but X-S.

Thus, the cost of shortages is inversely proportional to the storage costs.

As the order quantity X grows, the inventory costs increase and the shortages costs decrease correspondingly.

5.4.6   Final Remarks

The three variants of the basic EOQ model assume that the demand is sufficiently constant. In practice this is not necessarily realistic. The model used in this course is based on a slightly statistically fluctuating demand, superimposed with a seasonal distribution. The results show, however, that the models provide sufficiently good results for this purpose.

However, there are a couple of publications dealing with more precise methods for the calculation of the basic EOQ model for fluctuating demand (Juneau and Coates 2001; Teng and Yang 2007).

5.5   Task—Board Review Meeting 2

5.5.1   Agenda

1. Interpretation of the results of the SCM1 simulation (first fiscal year) for each area; SWOT analysis.
2. Selection and justification for the forecasting method(s) for each area.
3. Selection and justification for the inventory management method for each area.
4. Selection of the following parameters for an automated SCM solution:

   a) Delivery time after order receipt.
   b) Order cycles.
   c) Minimum inventory (incl. Shortages, if required).

5. Additional requirements for an automated SCM solution.

The calculations based on the results of the first fiscal year (Game 1) for the retail and distribution company of a company in comparison. The parameter for the selection of the optimal model is the minimum annual storage costs. For the retailer, this results in the shortage model and for the distributor almost the non-instantaneous model as the most cost-effective variants (Table 5.1).

Table 5.1 Typical results of three EOQ models

		Retailer			Distributor		
Results		Basic	Non-instant	Shortage	Basic	Non-instant	Shortage
Optimal order quantity [hl]	x_{opt}	2183.19	2447.93	2673.85	2160.67	2416.30	2415.70
Maximal inventory [hl]	M_l	2183.19	1947.08	1782.57	2160.67	1932.08	1932.56
Minimal inventory [hl]	l_{min}	2291.50	327.36	−563.93	320.64	320.64	−162.50
Minimal inventory costs [hl]	K_{min}	**1091.60**	**973.54**	**891.28**	**1080.33**	**966.04**	**966.28**
No. orders per year [hl]	A_b	27.29	24.34	22.28	27.01	24.15	24.16
Annual demand [hl]		59,579.00	59,579.00	59,579.00	58,356.00	58,356.00	58,356.00
Optimal backorder [hl]				891.28			483.14

References

Andler K (1929) Rationalisierung der Fabrikation und optimale Losgrösse. R. Oldenbourg, München

Beer A (2014) Der Bullwhip-Effekt in einem komplexen Produktionsnetzwerk. Springer Gabler, Wiesbaden

Chen F et al (2000) The impact of exponential smoothing forecasts on the bullwhip effect. Naval Res Logistics 47(4):269–286

Cronson R, Donohue K (2006) Behavioral causes of the bullwhip effect and the observed value of inventory information. Science 52(3):323–336

DecisionCraft (2010) Choosing the right forecasting technique. DecisionCraft Inc. http://www.decisioncraft.com/dmdirect/forecastingtechnique.htm. Accessed 16 Nov 2014

Dietl H (2012) Operations management. Universität Zürich. http://www.business.uzh.ch/professorships/som/stu/Teaching/F2012/BA/BWL/5_Lagerhaltungsmanagement.pdf. Accessed 16 Nov 2014

Harris F (1913) How many parts to make at once factory. Magazine Manag 10(2):135–136. 152

Holland W, Sodhi M (2004) Quantifying the effect of batch size and order errors on the bullwhip effect using simulation. Int J Logistics Res Appl 7(3):251–261

Juneau J, Coates E (2001) An economic order quantity model for time-varying demand. .Int J Mod Eng http://www.ijme.us/issues/spring2001/articles/economicorder.htm. Accessed 16 Nov 2014

Lee HL et al (1997) The bullwhip effect in supply chains. Sloan Manag Rev 38(3):93–102

Metters R (1997) Quantifying the bullwhip effect in supply chains. J Oper Manag 15(2):89–100

Sucky E (2009) The bullwhip effect in supply chains-an overestimated problem? Int J Prod Econ 118(1):311–322

Teng J, Yang H (2007) Deterministic inventory lot-size models with time-varying demand and cost under generalized holding costs. Inf Manag Sci 18(2):113–125. http://ijims.ms.tku.edu.tw/PDF/M18N22.pdf. Accessed 16 Nov 2014

Waser B (2010) Hochschule Luzern

Chapter 6
Game Round 2: Supply Chain Optimized

Abstract In this game round the supply chain will be optimized by integration of forecasting and inventory management. The teams select the necessary methods and parameters and a fiscal year is played in a partially automated environment. The results are reviewed and presented in a third review meeting. The results of the review meeting will be integrated into the system and the supply chain will be fully automated.

6.1 Game Rules

The game rules correspond mainly to those of game round 1 with the following differences:

1. All divisions (retailer, distributor, wholesaler, brewery) have complete transparency over the entire supply chain. The business divisions are correspondingly deciding on their order behavior (Fig. 6.1).
2. A demand forecasting method is for all divisions implemented and accessible by all divisions (Fig. 6.2).

© Springer-Verlag GmbH Germany 2017   41
K.-D. Gronwald, *Integrated Business Information Systems*,
DOI 10.1007/978-3-662-53291-1_6

	Retailer		Distributor		Wholesaler		Factory

Green Retailer		**Green Distributor**		**Green Wholesaler**		**Green Factory**	
Week	3	Week	3	Week	3	Week	3
Order In:	2'294 hl	Order In:	0 hl	Order In:	0 hl	Order In:	0 hl
Delivery In:	2'000 hl	Delivery In:	0 hl	Delivery In:	0 hl	Delivery In:	0 hl
Inventory:	2'347 hl	Inventory:	3'533 hl	Inventory:	5'518 hl	Inventory:	4'577 hl
Delivery out:	2'294 hl	Delivery out:	0 hl	Delivery out:	0 hl	Delivery out:	0 hl
Order out:	0 hl	Order out:	0 hl	Order out:	0 hl	Order out:	0 hl
Backorder:	0 hl	Backorder:	0 hl	Backorder:	0 hl	Backorder:	0 hl
Inventory cost:	$11'377	Inventory cost:	$11'691	Inventory cost:	$11'635	Inventory cost:	$8'342
Backorder cost:	$0	Backorder cost:	$0	Backorder cost:	$0	Backorder cost:	$0
I&B Cost:	$11'377	I&B Cost:	$11'691	I&B Cost:	$11'635	I&B Cost:	$8'342

Fig. 6.1 SCM2 game cockpit

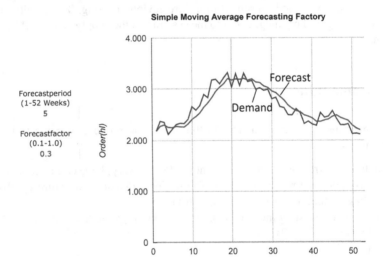

Fig. 6.2 Demand forecasting

6.2   Results

Typical results of this game round show that the Bullwhip effect could be avoided by introducing forecasting (Fig. 6.3).

Comparison with the previous year (game round 1) (Fig. 6.4):

The high level of inventory resulting from the first game was quickly reduced and maintained at an optimum level by consistent implementation of inventory management (Fig. 6.5).

Comparison with previous year (game round 1) (Fig. 6.6):

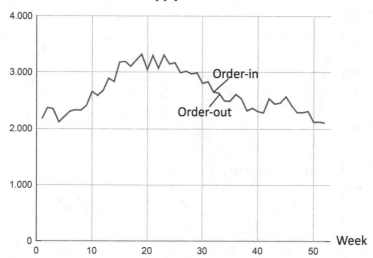

Fig. 6.3 Results factory with forecasting

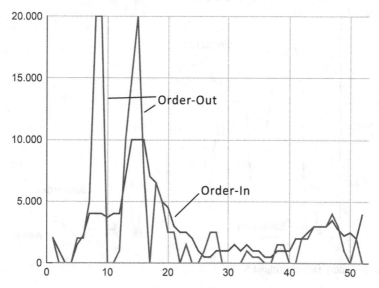

Fig. 6.4 Results factory without forecasting

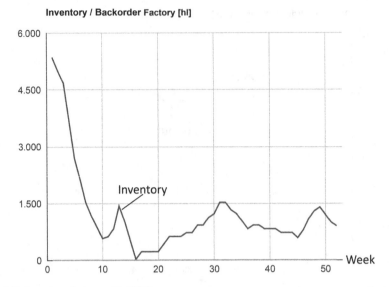

Fig. 6.5 Inventory factory with SCM

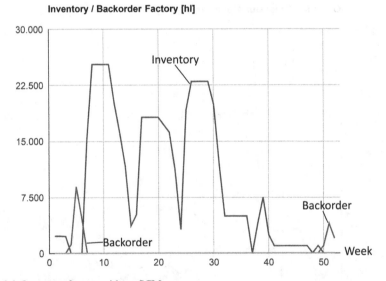

Fig. 6.6 Inventory factory without SCM

6.3   Task—Board Review Meeting 3

The results of the SCM integration are interpreted and forecasting and inventory management will be optimized.

The product portfolio can also be revised once again, especially regarding the portfolios of the competitors. The portfolios are visible to everyone now.

Chapter 7
Development and Implementation of a CRM Strategy

Abstract Customer relationship management is introduced as a customer oriented concept for the implementation of marketing strategies. The four methods of a CRM system (strategic, analytical, operational and communicative) are discussed and prepared for the SCM–CRM integration.

7.1 Initial Position

Customer Relationship Management is (like ERP and SCM) an entrepreneurial concept that has developed from a *tactical* marketing tool into a *strategic* element in all marketing decisions since the mid 1990s (Kumar and Reinartz 2012). As a *front-end system*, CRM is an integrated component of supply chain management systems and directly affects the demand with marketing campaigns. Forecasting is the link between CRM as a demand generator and SCM as a method for demand fulfillment. Forecasting automatically becomes a causal method. Instead of the time-series procedures used so far, CRM forces the use of cause-effect methods hereinafter.

The change from a *target-group oriented* to a *customer-centric* marketing is manifested in the evolutionary development of CRM. This change is taking place parallel to the evolution from the *product-oriented* to the *customer-oriented* entrepreneurial thinking and acting. Business Intelligence (BI) and Big Data Analytics (Big Data) are the *technology-driven* methods that support this development. CRM, BI and Big Data are increasingly interwoven and the boundaries are flowing, as for example in *Analytical CRM*. After completion of the company-internal optimization tasks ERP and SCM, this customer-oriented evolution will dominate as this course evolves.

Kumar and Reinartz (2012) identified four generations of CRM methods between 1990 and 2008:

- **Functional CRM (first generation)** that integrates the independent methods *salesforce automation (SFA)* and *customer Service and Support (CSS)*. SFA encompasses all activities in internal and external sales. CSS deals with after-sales activities such as help desk, call center, and field service support.
- **CRM—customer-oriented front-end (second generation)** bundled all activities with customers to a unified view regardless of the goals of the customer

© Springer-Verlag GmbH Germany 2017
K.-D. Gronwald, *Integrated Business Information Systems*,
DOI 10.1007/978-3-662-53291-1_7

contact (presales, sales, post sales). This approach, which was primarily technological and tool-oriented, led to a disillusionment in the use of CRM systems in the mid-1990s. Customer expectations exceeded the possibilities of these systems by far. This led to a rethinking of CRM from an IT tool to CRM as an entrepreneurial concept with a more strategic viewpoint.

- **CRM—strategic approach (third generation)** is the paradigm shift of the CRM objectives from cost control to generate sales and growth. This led to the integration of CRM as a front-end system with back-end systems such as ERP and SCM.
- **Agile and flexible CRM strategy (fourth generation)** consequently exploits the possibilities of the internet age such as cloud computing, social media, web-based services, self-services. These opportunities open CRM as a strategic tool for small and medium-sized enterprises without any major investment in technology and tools.

7.2　CRM: Objectives and Methods

Customer Relationship Management refers to practices, strategies and technologies to manage customer relationships with the primary objectives of customer acquisition, the expansion of the customer base, customer selection, the identification of profitable customers and customer loyalty, an effort to keep existing customers (Schmid and Bach 2000).

Regarding the objectives and procedural models, today four basic methods are common: *Strategic CRM, Analytical CRM, Operative CRM, Communicative CRM*.

7.2.1　Strategic CRM

The goal of *Strategic CRM* is to build as much knowledge as possible about customers, to use this knowledge to optimize the interaction between companies and customers, with the aim of maximizing the *Customer Lifetime Value (CLV)* for the company (Kumar and Reinartz 2012).

7.2.2　Analytical CRM

Analytical CRM uses customer data to form profitable relationships between customers and companies. It uses traditional business intelligence (BI) methods such as data warehouse, data mining, and online analytical processing systems (OLAP) to determine customer satisfaction and active measures to optimize the corresponding parameters. In this course, the customer loyalty plays a particular role. This can be measured directly through the *share of wallet*. It means how much per cent of its beer consumption a customer shares between the four brands. Marketing measures can be derived from this (Operative CRM).

7.2.3 Operative CRM

Operative CRM implements the identified measures of strategic CRM which were quantified in analytical CRM in the form of (automated) solutions for marketing, sales and services. This course deals exclusively with *campaign management* as a basis for appropriate marketing campaigns.

7.2.4 Communicative CRM

Communicative CRM includes the management of all communication channels between the customer and the company (telephony, internet, e-mail, direct mailing, etc.). The various communication channels are synchronized, controlled and targeted to enable bidirectional communication between customers and companies. This approach is also referred to as *multichannel management* (Grabner-Kräuter and Schwarz-Musch 2009, p. 184).

7.3 Task: Board Review Meeting 4

Board review meeting 4 serves as preparation for the next game round.

The teams analyze their market position and competition situation. They analyze the share of wallet with key accounts, evaluate their product strategy and develop a key account strategy.

Templates for each team are available as downloads from the download area in the visitor center.

References

Grabner-Kräuter S, Schwarz-Musch A (2009) CRM Grundlagen und Erfolgsfaktoren. In: Hinterhuber H, Matzler K (Hrsg) Kundenorientierte Unternehmensführung, Kundenorientierung – Kundenzufriedenheit – Kundenbindung, 6. Auflage. Gabler, Wiesbaden, pp 174–189

Kumar V, Reinartz W (2012) Customer relationship management concept, strategy, and tools. Springer, Berlin

Schmid E, Bach V (2000) Customer relationship bei Banken. Bericht Nr. BE HSG/CC BKM/4. Universität St. Gallen

Chapter 8
Game Round 3: CRM–SCM-Integration

Abstract After the automation of the supply chain, the four companies compete directly. Demand generation dominates and is integrated with demand forecasting and inventory management. The game is managed on a quarterly basis with the help of promotions, pricing and portfolio management. Lack of business intelligence makes a proactive action impossible for the companies and the situation is similar to that of the non-communicative supply chain in the first round, now only in relation to the entire market.

8.1 Initial Situation

Game round 3 differs significantly from the first two. The supply chain is automated and orders and deliveries are executed automatically. The four teams are in direct competition with each other and their measures influence each other. Decisions are made monthly. There are 12 rounds. There is an initial budget of 5,000,000 ¤ (¤ your local currency). Supervisor can replenish the marketing budget any time during the game in appropriate. Teams analyze their results each month. Students can perform the following marketing activities:

8.2 CRM: Promotion for One or More Products

Promotions are done for each individual product and each individual retail chain. A promotion costs 250,000 ¤ for each product and the specific retailer increasing the market share for this product and this retailer permanently by 30%. The corresponding market shares of the three competitors are reduced by 5% accordingly, 15% are taken from the beer giants. The lack of transparency between the competitors in the absence of business intelligence can result in the situation that one or more other competitors make a promotion for the same product at the same time which will (partly) compensate expected market growth and has impact on EBITDA. Price and market agreements between competitors are of course prohibited. The supply chain management systems adjust demand changes automatically.

© Springer-Verlag GmbH Germany 2017   49
K.-D. Gronwald, *Integrated Business Information Systems*,
DOI 10.1007/978-3-662-53291-1_8

8.3 CRM: Price Change for One or More Products

A price change (price increase, price decrease) can be done for individual products in individual retail chains. A price change costs 200,000 ¤ and results in a change of market share by 10% in both directions (price increase: −10%, price reduction: +10%). The impact on the competitors will be 5% each in both directions accordingly (price increase at e.g. Alpha: +5% for Green Beer, Royal Beer, Wild Horse Beer. Price decrease at e.g. Alpha Beer: −5% for Green Beer, Royal Beer, Wild Horse Beer).

8.4 CRM: Discount for One or More Products

A discount has similar effects like a promotion, just for a limited time of one cycle (1 month). It costs 200,000 ¤ and results in an effective annual increase by 10% and a corresponding reduction by 5% for the other competitors.

8.5 CRM: Change of Product Portfolios

The change of product portfolios is deactivated by default, but can be activated by the supervisor if needed. Taking a product out of the portfolio will reduce revenue to zero from the month it was taken out onwards resulting in a corresponding increase in market share of 3% each for that product at the competitors. Taking a new product into the portfolio will result in an initial market share of 20% for the new product and a decrease of 5% each for the competitors for the corresponding product.

8.6 CRM: Share of Wallet at Key Accounts

New in this round is the sales to four independent large retail groups (KDISCOUNT, KDISUPER, KDIvalue, KDIFRESH). Direct business with the own four retail chains contributes to just about 30% of the total revenue. About 70% of the revenue comes from the four key accounts. Proper managing those accounts has significant impact on the market share. There are two perspectives of the share of wallet, one for the revenue distribution between the four key accounts at Alpha Beer, Green Beer, Royal Beer and Wild Horse Beer and one for the revenue distribution of Alpha Beer, Green Beer, Royal Beer, Wild Horse Beer at each of the four key accounts. The share of wallet can be influenced by increasing or decreasing the discount for the corresponding key account. Discount increase and decrease will be done with increments of 2% in both directions. An increase of discount by 2% increases revenue by 20%, a decrease in discount reduces revenue by 20% at the respective key account. It has no direct impact on the other teams and does not affect marketing budget.

Chapter 9
Business Intelligence (BI) and Big Data Analytics (Big Data)

Abstract Business Intelligence and Big Data Analytics are separated from one another and introduced as independent complementary methods. Predictive analytics, sentiment analysis and social media analytics are introduced as Big Data methods for active market management on an individual customer basis and prepare the last game round of a transparent and proactive market influence with realtime intelligence. Theoretical foundations and algorithms as well as a comprehensive treatment of social media text mining and statistics are developed in Part II, Chap. 14 parallel to this chapter. The game round 4 and a final analysis will finish this course.

9.1 Initial Situation

The planning and execution of marketing activities during game round 3 was limited by the fact that there was no information about the market behavior of the other teams. The impact of the competitors' decisions was only available as a growth in sales or a decrease in one team's own numbers after processing 1 month. Likewise, the information about which companies were responsible for the results was missing. There was a similar situation in game round 1 due to lack of transparency in the supply chain. Now it was the lack of transparency about the market. This affected the behavior of the competitors as well as external influences, which made the forecasting more difficult. Business intelligence (BI) and big data analytics can fill this gap with *data mining* and *predictive analytics*.

9.2 BI and Big Data Separation

In Part II, BI is still treated as separate sections to explain traditional methods such as OLAP and ETL, but more for historical reasons. With the increasing use of in-memory databases (Bayer 2013), the separation of OLAP and OLTP is outdated and the ETL process becomes obsolete (Plattner 2013). Data mining, however, is fully integrated into big data projects. Big data analytics also uses business intelligence methods and processes to a large extent.

© Springer-Verlag GmbH Germany 2017
K.-D. Gronwald, *Integrated Business Information Systems*,
DOI 10.1007/978-3-662-53291-1_9

Analytical CRM uses business intelligence methods for target group oriented marketing. The paradigm shift from CRM to customer-oriented thinking and acting is best illustrated by the distinction between BI and Big Data.

Traditional data mining addresses the segmentation of customers into groups for target-group-oriented marketing (income, social status, place of residence, education, gender, age, …). It tries to identify parameters that determine optimal customer groups and includes the development of models to assign customers to these target groups. The customer does not exist as an individual, as a person, in these models.

Big Data is the step from product oriented to the customer oriented cross-selling and up-selling with real-time analysis of the customer behavior on an individual basis with the focus on individual customers and not on generic markets.

Big data analytics will be discussed in detail in Part II, Chap. 14. At this point the focus is on eliminating weaknesses that have arisen during the CRM game. This section is therefore limited to the issues relevant to the solution of these problems.

9.3   Analytics Evolution

9.3.1   Descriptive Analytics

Descriptive analytics is the basic form of analysis. It is past-related. Descriptive analytics still dominates the entire business analytics. It answers questions about what happened and why it happened and examines historical data using data mining in terms of factors for success or failure. Most management reports for sales, marketing, operations and finance use this type of *post mortem analysis* (Bertolucci 2013).

9.3.2   Predictive Analytics

Predictive Analytics is already future oriented. It can forecast what might happen in the future. Here, historical data is combined with rules, algorithms, and sometimes external data to make predictions about e.g. market developments. It uses comprehensive statistical methods, models, data mining and machine learning to analyze current and historical data and to derive predictions for the future (Bertolucci 2013).

9.3.3 Prescriptive Analytics

Prescriptive Analytics describes not only what and when an event can occur, but why it can happen. Prescriptive Analytics makes proposals for decision-making options to take advantage of the future or to minimize risks. It uses hybrid data, combining structured (numbers, categories, …) and unstructured data (videos, images, sound, text, …) as well as business rules to make data-oriented decisions without compromising other priorities. Prescriptive Analytics can dynamically and continuously analyze data real-time and thus continuously update predictions (Bertolucci 2013).

9.3.4 Sentiment Analysis

Sentiment analysis is a common type of predictive analysis. It can analyze emotions, opinions, assessments and attitudes of individuals versus organizations, products, services, other people, subjects, events, questions, how people communicate via social media, text, video, and other online media. These communications fall into three basic categories: positive, neutral or negative. There are a couple of other terms with slightly different goals for this type of analysis: *sentiment analysis, opinion mining, opinion extraction, sentiment mining, subjectivity analysis, emotion analysis, review analysis*. Sentiment Analysis is located between descriptive and predictive analytics, depending on the author (Bertolucci 2013; Sharef 2014).

9.3.5 Text Mining

Text mining is primarily the method of processing unstructured text in such a way that it can be further treated with other analytical methods to gain information (Hardoon and Shmueli 2013). Text analytics includes the processes

- Content Categorization: Classification of text documents into categories.
- Text Mining: Recognizing patterns and structures and making predictions or understanding the behavior.
- Sentiment Analysis: Assessment of text content as positive or negative (polarization)

There are two basic approaches to text mining, the linguistic approach, the attempt to determine structure and meaning through grammatical rules, and the mathematical approach of numerical methods to extract as much information from texts as possible. The mathematical approach requires several steps to transform text data into a numerical form, which is understood by mathematical, analytical methods. Sentiment analysis is understood as a linguistic approach (Duffy 2008).

Typical applications for text mining are search engines, email spam filters, fraud detection, customer relationship management, social media analysis, marketing studies, web content analysis, ...

9.4   Task: Board Review Meeting 5

As preparation for the final round, students will prepare a final board review meeting. They will analyze their CRM result (SWOT analysis, and plan the use of business intelligence and big data analytics.

9.5   Game Round 4: CRM–Big Data Integration

The rules of this round correspond mainly to those of game round 3. In addition, big data for prescriptive analytics is now introduced. Social media text mining and sentiment analysis are considered as methods. This data comes from the web real-time and is integrated into the game. In addition, virtual data of the four companies are inserted and thus reality and virtual world are networked with each other. This information helps the teams in their decisions. The results of sentiment analysis can have a direct impact on the forecast and the results, so they require action by the management of the four companies.

References

Bayer M (2013) Hadoop – der kleine Elefant für die grossen Daten. Computerwoche. http://www. computerwoche.de/a/hadoop-der-kleine-elefant-fuer-die-grossen-daten,2507037.   Accessed 16 Nov 2014

Bertolucci J (2013) Big data analytics: descriptive vs. predictive vs. prescriptive. Information Week. http://www.informationweek.com/big-data/big-data-analytics/big-data-analytics-descrip tive-vs-predictive-vs-prescriptive/d/d-id/1113279. Accessed 16 Nov 2014

Duffy V (2008) Handbook of digital human modeling: research for applied ergonomics and human factors engineering. Taylor & Francis. http://books.google.ch/books?id=Ira9qiakiTMC. Accessed 16 Nov 2014

Hardoon D, Shmueli G (2013) Getting started with business analytics: insightful decision-making. Chapman & Hall/CRC machine learning & pattern recognition series

Plattner H (2013) A course in in-memory data management the inner mechanics of in-memory databases. Springer, Berlin

Sharef N (2014) A review of sentiment analysis approaches in big data era. Universiti Putra Malaysia. http://www.academia.edu/8716357/A_Review_of_Sentiment_Analysis_Approaches_ in_Big_Data_Era. Accessed 16 Nov 2014

Part II
Theory and Background

Abstract

This part contains the theoretical background as well as the thematic and formal deepening of the individual components of an integrated business information system environment ERP, SCM, CRM, BI, Big Data Analytics.

Chapter 10
ERP: Enterprise Resource Planning

Abstract The ERP part deals extensively with the strategic objectives of an ERP implementation. It is complemented by topics such as ERP template development and rollouts, total cost of ownership and organizational change management. A comprehensive chapter on organizational readiness deals with best practices quality and process models complemented by a detailed treatment of post-merger IT integration.

10.1 Definition and Goals

Sales structures, business processes as well as the IT infrastructure are linked via ERP and can thus be standardized using a suitable ERP strategy and implemented with suitable software systems. The operational view of an ERP system describes it *as a system that supports all business processes running in a company. It contains modules for procurement, production, sales, equipment management, finance and accounting etc., via a common Data base* (Springer Gabler Verlag [1]).

The inclusion of business objectives in the considerations leads to a more strategic approach to the implementation goals of an ERP system. *The standardization of business processes beyond organizational boundaries can have enormous synergy effects. Organizations can implement best practices in the system, and the ERP system is perceived as a business tool rather than as an IT tool* (Desai and Srivastava 2013).

10.2 Strategic Objectives

The three fundamental business objectives of an ERP implementation are:

- Creation of a common business process architecture
- Standardization of internal and external master data
- Standardization of the information system architecture

© Springer-Verlag GmbH Germany 2017   59
K.-D. Gronwald, *Integrated Business Information Systems*,
DOI 10.1007/978-3-662-53291-1_10

10.2.1 Standardization of Business Processes

A business process is a collection of linked tasks in a value chains with one or more inputs and a customer output (Springer Gabler Verlag [2]). Business processes are not bound to organizational boundaries; rather, a couple of internal departments, as well as business-specific business partners and resources, can be integrated (Bächle and Kolb 2012).

Standardization of business processes means creating a uniform and integrated process landscape in a company or between companies to be able to control the exchange of services between business units, as well as with external customer suppliers or partners, transparently and efficiently (Schmelzer and Sesselmann 2008).

At merger & acquisition, process standardization supports the implementation of strategic goals and the creation of a uniform corporate culture. However, this is usually associated with power shifts on management level.

In addition, process standardization facilitates the rapid and company-wide implementation of process improvements (best practice sharing), the utilization of synergies, creates uniform company interfaces with customers, suppliers and partners and creates the prerequisites for bundling or outsourcing business processes (Schmelzer and Sesselmann 2008).

10.2.2 Localization

Not all business processes can be standardized globally. By consistently implementing a global business process standardization, localizations can be reduced to 10–20% of the total business processes. Back office functions such as purchasing, accounting, supplier management, etc. are easy to standardize. In the CRM area, local influences of consumer behavior and cultural influences play different roles. Especially from local branches, adherence to conventional structures (resistance against change) is often justified by local peculiarities. This is, however, to be viewed in a more differentiated manner and constantly changing within the framework of globalization.

This can be best illustrated with one of the world's most standardized products, MacDonald's Big Mac and the countries of India and China. India and China have both emerged into industry nations with a strong purchasing power and a globalized middle class. MacDonald's is very successful in both countries, however, with very different localizations.

India is a country with a regionally diverse but stringent eating culture, which is influenced by the dominant religions Hinduism and Islam. In India, MacDonald's is very much focused on local cultural characteristics. India is the only country in the world where McDonald's serves neither beef nor pork in its products, but at the same time has a wide variety of vegetarian specialties in its portfolio. Marketing and branching are *indianized* with a high acceptance as a fast food restaurant (Mathur 2011).

China, on the other hand, has widely adopted the American MacDonald's culture and the society has evolved in a direction that fits MacDonald's culture (singles, one-child small families). The cultural adaptation took place in both directions, on the one hand through the acceptance of the customers by disciplined waiting in the queue and self-seating, on the other by McDonald's change from a fast food restaurant to a leisure center for seniors and students. Social aspects and permanent evolution in (Chinese) society are of great importance, which is constantly changing (Watson 2000).

CRM and big data analytics with the focus on individual customer needs play a significant role for the rapid collection of social trends.

Reasons for localization include country-specific tax regulations, legal regulations, wage and salary systems, language-specific requirements, and the local marketing and branding requirements described above.

Areas where localizations can occur are jobs, organization, processes, functions, data and technology.

The art is the balance between *desires* and *necessities*.

10.2.3   Master Data Standardization

The standardization of master data is prerequisite for the centralization of purchasing. Master data are the basic data of a company. This includes article numbers, customer data, employee data, vendor list, parts lists...

According to Heutschi et al. (2004) and Schemm (2008), the following types of standards can be identified:

Format standard unify the syntactic encoding of data by specifying the sequence, length, and type of data elements.

Data standards unify structure and semantic coding of data by specifying the data elements.

Message/document standards serve to unify the aggregation or linking of data elements to messages for transmission between information systems.

Process standards unify organizational processes by defining the dependencies between individual tasks.

Business standards create a uniform legal framework.

The *data quality* is another prerequisite for successful master data optimization. For companies without master data management, 50% or more of the master data can be obsolete or redundant (Johnson 2005). The most important data quality dimensions (Schmidt 2010) are

- Accuracy
- Completeness
- Actuality

- Consistency
- Redundancy free
- Accessibility

In addition to the optimization of the order-to-cash process within the supply chain management, master data standardization contributes significantly to cost savings in ERP implementations. The global harmonization of the product numbers for raw and finished products is the prerequisite for the centralization of purchasing and allows for the negotiation of more favorable purchasing conditions and contracts with suppliers. It improves the ability to deliver and optimize inventory in globally networked supply chains (Johnson 2005).

10.2.4   Standardization of IT Infrastructure

The standardization of the IT infrastructure includes the standardization of hardware and software, including the agreements for global purchasing conditions and maintenance agreements with suppliers, the consolidation of global data centers and the decision for a standard ERP software. Separate packages for CRM and SCM are optional and depend on the requirements and the situation.

For the selection and implementation of standard software packages (ERP, SCM, CRM) a principal decision must be taken for an in-house solution (on premise) or for a cloud solution (SaaS). A key parameter for one or the other alternative is the *total cost of ownership (TCO)*.

10.2.5   TCO: Total Cost of Ownership

The term "total cost of ownership" describes the total costs of investments (computer systems) during their lifecycle in the company.

The total costs are divided into acquisition (hardware, software), operation (server, networks), technical support (maintenance, user training, assistance) and user activities. The most important basis for the understanding of TCO is the distinction between *direct* and *indirect costs*.

Direct costs (approx. 60%) are incurred as capital costs, administration, technical support. Fixed costs of operation and maintenance independent of the provision of services (HW, software, network operation, security, services, operation helpdesk) are possible. Direct costs are budgetable.

Indirect costs (approx. 40%) are incurred by end users and system operations. They can be influenced by the quality of the end user support and are difficult to budget (training, availability of the systems, response times of the systems).

10.2.6 ERP-Template

The basic idea of an ERP system is a set of standardized business process modules that can be adjusted to (almost) any real business process situation without programming by configuring parameters and master data. These modules are arranged around a central fixed hardware software core (server database).

In reality, it is often not possible to map all customer processes with standard configurations. There are organizational, financial, legal and company policy reasons that prevent complex business process reengineering.

Basically there are three possible solutions:

a) The company adapts its business processes to the ERP standard (organizational change management).
b) Additional functions must be programmed (customization).
c) A combination of a) and b).

These additional programs may cause significant additional costs. In the case of release changes, the follow-up costs may be even higher than the costs for a complete new implementation. They are often written by external companies that do not necessarily adhere to prescribed standards and thus do not generate upwards compatible code if the manufacturer of the ERP system changes its standards. Within the framework of service contracts, the upward compatibility of standard modules is normally guaranteed when there is a release update. External programs are usually excluded. This also applies to those that have been written with a programming language provided by the ERP vendor.

The terms *configuration* and *customization* are used as follows:

Configuration is the representation of the value chain of a company in an ERP system by exclusive use of standardized business process modules of the system without external programming.

Customization is the addition or modification of business process modules by external programs which are not available in the standard of the ERP system. This includes interface programs to third party systems and reports.

These definitions apply to SCM and CRM systems accordingly.

An *ERP template* is the representation of the value chain of a company in which 80–90% of all business processes are standardized as a combination of *configuration* and *customization* at all levels (divisions, business units, subcontractors, branches). Centrally developed, maintained and rolled out (Fig. 10.1).

The rollout of a template based ERP system is complex and involves considerable risks, especially at the organizational level. It affects organizational boundaries, management, working habits, organizational structures and decision making processes, forcing new forms of collaboration within and between business units. Employees will use new tools and other information, learn new skills, collaborate with other people in new relationships, learn to deal with new responsibilities and new metrics to assess their performance.

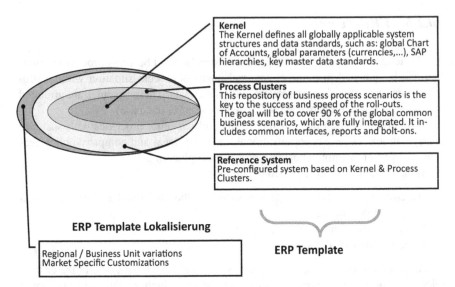

Fig. 10.1 ERP template

These changes create uncertainty among those affected. Insufficient acceptance (*resistance against change*), caused by inadequate training, new and modernized processes, shifting power positions, ... requires a comprehensive *organizational change management*. These organizational adjustments often represent a greater project risk than the associated technical changes.

The prerequisite for the successful rollout of an ERP template is *organizational readiness*. *Organizational change management* is the formal process that leads to *organizational readiness*.

10.2.7   Organizational Change Management

Implementation of organizational change management requires three steps:

- Preparation
- Implementation
- Integration into the organization

Objectives of Organizational Change Management

- Increase *readiness* of end users and organization during preparation.
- Increase the speed with which the affected organizational areas adapt to the new situation while minimizing the risks of interruptions and losses of productivity during the implementation.
- Integration into the organization while at the same time achieving the business objectives.

A structured approach to minimize risks while achieving business objectives includes the following steps

Achievement of Business Objectives

- Focus on the business objectives and not on the technical system installation.
- Integration of processes and technology with people and organization.

Promotion of Leadership

- Link the success of the program with the success of the executives.
- Development of a new management behavior by managers.

Development of a Change Vision

- Development of a change readiness, a willingness to change.

Definition of a Change Strategy

- Implementation of an implementation path that achieves rapid success.
- Provide a toolkit for the implementation of the three phases (preparation, implementation, anchoring).
- Definition and implementation of a communication plan.
- Definition and implementation of a training plan.
- Definition and implementation of a recognition and reward strategy.

Manage Employee Performance

- Recognition and rewarding new skills and competences.
- Selection and implementation of new performance evaluations to enhance new behavior.

Development of a New Corporate Culture

- Identification of necessary cultural adaptations for the new ways of cooperation.
- Observation of individual behavior regarding the new forms of cooperation.

Design of the New Organization

- Definition of activities, roles and responsibilities.
- Identification of the changes and their influence on the organization.

10.2.8   Localization Requirements for a Template

Necessary localization requirements are already identified, assessed and either rejected or implemented during template development (Fig. 10.2).

The localization activities in Fig. 10.2 are listed with increasing effort and associated costs. The one on top, modification of the original program code in a standard software package, is only listed because there have been companies that have done just that. This is a time bomb and, of course, strictly forbidden. A viable,

Fig. 10.2 ERP template localization

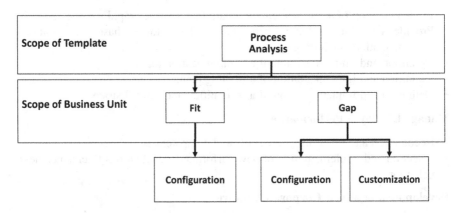

Fig. 10.3 ERP template localization process analysis

albeit expensive, way, the software supplier is being asked by customers to add this as a new function to the program. Software suppliers sometimes do this for their key accounts or when the new solution fills a market gap and fits well into the existing package.

Managing localization requirements for a template requires corresponding processes and organizational structures (*governance*) during template development, rollout and later operation.

The first step is a fit-gap analysis. This checks to what extent the corresponding process is already covered by standard functions of the template. Those parts of the business process or value chain, which are already covered by the template, are included directly in the configuration requirements for the local rollout. A gap is then analyzed to which extent these can be fulfilled with standard functions (*configuration*) and which parts must be supplemented by external programs (*customization*). The local process is then included in the template (Fig. 10.3).

Before this happens, however, the process has to go through an approval process (*change control*). Localizations are significant threats for the stability of a template.

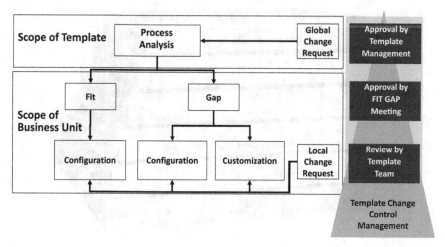

Fig. 10.4 Template localization governance

It must be protected by appropriate formal processes. In this example, it is a procedure with three levels. In addition, there is the distinction between a local change request that affects only one business unit and a global change request that affects the entire template (Fig. 10.4).

10.3 Organizational Readiness

ERP template rollouts require a significant amount of organizational and behavioral change of people and organizations. That requires that organizations and people are *changeable*. Organizational readiness is the measure for the *changeability* of an organization. *Organizational change management* is the method to eliminate *resistance against change* and improve *organizational readiness*. The degree of organizational readiness can be measured and controlled by a set of established quality and process models (Fig. 10.5).

For IT organizations, *ITIL (IT Infrastructure Library)* (www.itilfoundation.org) is the de facto standard for the transformation of an IT organization from a technology driven administration to an IT service provider. Business and IT services are managed by *service level agreements (SLAs)* at ITIL.

On the other hand, *CMMI (Capability Maturity Model Integration)* (www.cmmiinstitute.com) has expanded from an initially IT-centric model to a wide range of corporate processes and now supports a cross-departmental process integration, which can be used as a measure for *organizational readiness*. (CMMI 2014).

Both process models are integrated into a network of other quality management systems for process improvement, such as *Six Sigma* (http://www.sixsigma-

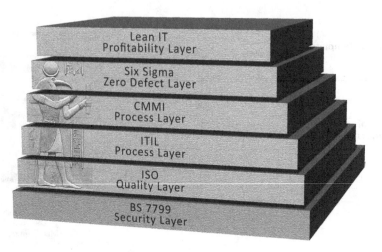

Fig. 10.5 IT quality and process models

institute.org) and *Lean IT* (www.lean.org), a prerequisite for outsourcing to global IT service providers.

10.4   BS7799 and ISO20000

The basis is the standard for information security management BS7799 (ISO 17799, ISO27001)   (http://iso-17799.safemode.org/index.php?page=BS7799-2).   The ISO9000 quality standard was for years the standard for quality management. It has been superseded by ISO20000, used today by many global IT service providers. It has become the standard for IT service management and is close to ITIL (ISO/IEC20000 2012).

10.5   ITIL: Information Technology Infrastructure Library

ITIL provides a framework of Best Practice guidance for IT Service Management and since its creation, ITIL has grown to become the most widely accepted approach to IT Service Management in the world (www.itilfoundation.org).

Core publications within ITIL (Figs. 10.6 and 10.7):

- Service Strategy
- Service Design
- Service Transition
- Service Operation
- Continual Service Improvement.

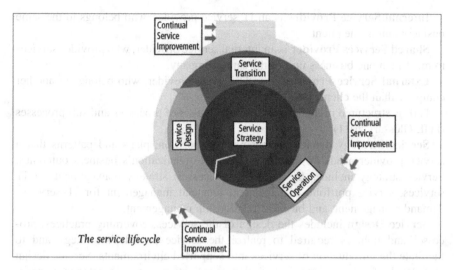

Fig. 10.6 The service lifecycle, source: www.itil.org

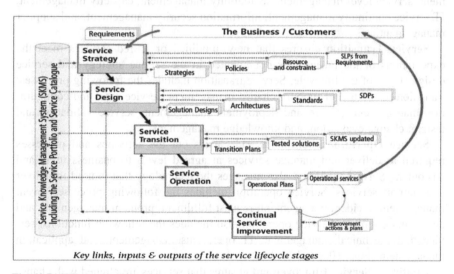

Fig. 10.7 ITIL service lifecycle stages, source: www.itil.org

With the increasing integration of business and IT ITIL has become the basis for the transformation of an IT organization from providing technology and infrastructure to a customer-oriented service provider.

The term *service provider* can refer to internal or external organizations. There are three different types of services providers (Bucksteeg 2012):

Internal Service Provider is an IT service provider, who belongs to the same business unit as the client.

Shared Services Provider is an internal service provider, who provides services to more than one business unit of the same company.

External Service Provider is an IT service provider, who belongs to another company than the client.

ITIL is structured in five phases, each with a set of processes and sub-processes (ITIL Glossary 2011):

Service Strategy defines the perspective, position, plans and patterns that a service provider needs to execute to meet an organization's business outcomes. Service strategy includes the following processes: strategy management for IT services, service portfolio management, financial management for IT services, demand management, and business relationship management.

Service Design includes the design of the services, governing practices, processes and policies required to realize the service provider's strategy and to facilitate the introduction of services into supported environments. Service design includes the following processes: design coordination, service catalogue management, service level management, availability management, capacity management, IT service continuity management, information security management, and supplier management.

Service Transition ensures that new, modified or retired services meet the expectations of the business as documented in the service strategy and service design stages of the lifecycle. Service transition includes the following processes: transition planning and support, change management, service asset and configuration management, release and deployment management, service validation and testing, change evaluation, and knowledge management.

Service Operation coordinates and carries out the activities and processes required to deliver and manage services at agreed levels to business users and customers. Service operation also manages the technology that is used to deliver and support services. Service operation includes the following processes: event management, incident management, request fulfilment, problem management, and access management. Service operation also includes the following functions: service desk, technical management, IT operations management, and application management (Figs. 10.8 and 10.9).

Continual Service Improvement ensures that services are aligned with changing business needs by identifying and implementing improvements to IT services that support business processes. The performance of the IT service provider is continually measured and improvements are made to processes, IT services and IT infrastructure to increase efficiency, effectiveness and cost effectiveness. Continual service improvement includes the seven-step improvement process.

Two ITIL processes overlap with CMMI:

- Service Level Management (phase Service Design)
- Improvement Process (phase Continual Service Improvement)

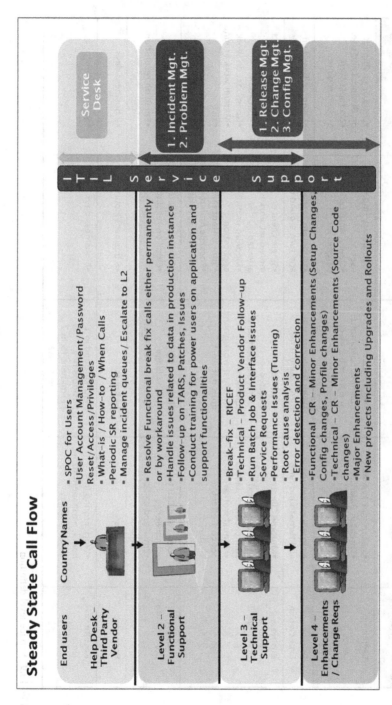

Fig. 10.8 ITIL service operation, source: www.itil.org

Service Level	Critical	Urgent	Normal	Low
Definition	▪System not available ▪Threaten Business Continuity	▪High impact on business operations or part of business operations	▪Temporary impact to user ▪Query, Application of patches etc	▪Little or no impact on Business and user ▪No rapid turn around required
Service Expectation	▪Immediate Response and turnaround time	▪High Response and turnaround time	▪Moderate response and turn around time	▪Taken up based on resource availability
Business Impact & Urgency	▪Multiple Clients are affected ▪Mission critical application impacted ▪Manual workaround is not cost effective	▪Few Clients are affected ▪Affecting particular function ▪Manual workaround does not exist	▪Single client or individual work is affected ▪Medium priority application ▪Intermittent failures ▪Possible manual workaround exist	▪Single client impact ▪Low priority application ▪No impact on work ▪Manual work around exist ▪Service request
Request Confirmation	▪Email and Call to acknowledge ▪Email and call on resolution	▪Email or Call to acknowledge ▪Email on resolution	▪Email to acknowledge ▪Email on resolution	▪Email to acknowledge ▪Email on resolution
Response Time	▪45 Min	▪4 hours	▪8 business Hours	▪2 business days
Resolution Time	▪Work around in 4 hours, permanent fix in 1 business day	▪Workaround in 1 business day, permanent fix in 2 business days	▪5 business days	▪Scheduled

Fig. 10.9 Service level definition

10.6 CMMI: Capability Maturity Model Integration

CMMI (Capability Maturity Model Integration) models are collections of best practices that help organizations to improve their processes. These models are developed by product teams with members from industry, government, and the Software Engineering Institute (SEI). A Capability Maturity Model (CMM), including CMMI, is a simplified representation of the world. CMMs contain the essential elements of effective processes (CMMI 2011).

This model, called CMMI for Services (CMMI-SVC), provides a comprehensive integrated set of guidelines for providing superior services. The CMMI-SVC model provides guidance for applying CMMI best practices in a service provider organization. Best practices in the model focus on activities for providing quality services to customers and end users. CMMI-SVC integrates bodies of knowledge that are essential for a service provider.

There are four *capability levels*. They are numbered from 0 to 3 (CMMI 2011) (Table 10.1):

0. Incomplete
1. Performed
2. Managed
3. Defined

A capability level for a process area is achieved when all generic goals are met at this level (CMMI 2011).

Additional, there are five levels. They are numbered from 1 to 5 (CMMI 2011) (Table 10.1):

1. Initial
2. Managed
3. Defined
4. Quantitatively Managed
5. Optimizing

A maturity level consists of specific and generic practices for a predefined set of process areas that improve the overall performance of an organization. The maturity level of an organization provides a way to describe its performance. The

Table 10.1 CMMI—comparison of capability and maturity levels

Level	Capability levels	Maturity levels
Level 0	Incomplete	
Level 1	Performed	Initial
Level 2	Managed	Managed
Level 3	Defined	Defined
Level 4		Quantitatively managed
Level 5		Optimizing

Source: CMMI (2011)

maturity level is measured by how the specific and generic goals of the predefined process areas are achieved. (CMMI 2011).

10.6.1   CMMI—Capability Level

10.6.1.1   Capability Level 0: Incomplete

An *incomplete process* is a process that either is not performed or is partially performed. One or more of the specific goals of the process area are not satisfied and no generic goals exist for this level since there is no reason to institutionalize a partially performed process.

10.6.1.2   Capability Level 1: Performed

A capability level 1 process is characterized as a *performed process*. A performed process is a process that accomplishes the needed work to produce work products; the specific goals of the process area are satisfied.

10.6.1.3   Capability Level 2: Managed

A capability level 2 process is characterized as a *managed process*. A managed process is a performed process that is planned and executed in accordance with policy; employs skilled people having adequate resources to produce controlled outputs; involves relevant stakeholders; is monitored, controlled, and reviewed; and is evaluated for adherence to its process description.

10.6.1.4   Capability Level 3: Defined

A capability level 3 process is characterized as a *defined process*. A defined process is a managed process that is tailored from the organization's set of standard processes according to the organization's tailoring guidelines; has a maintained process description; and contributes process related assets to the organizational process assets.

10.6.2   CMMI—Maturity Level

A maturity level consists of related specific and generic practices for a predefined set of process areas that improve the organization's overall performance. The maturity level of an organization provides a way to characterize its performance. A maturity level is a defined evolutionary plateau for organizational process improvement. Each maturity level matures an important subset of the organization's processes, preparing it to move to the next maturity level.

10.6.2.1   Maturity Level 1: Initial

At maturity level 1, processes are usually ad hoc and chaotic. The organization usually does not provide a stable environment to support processes. Success in these organizations depends on the competence and heroics of the people in the organization and not on the use of proven processes. Despite this chaos, maturity level 1 organizations provide services that often work, but they frequently exceed the budget and schedule documented in their plans.

10.6.2.2   Maturity Level 2: Managed

At maturity level 2, work groups establish the foundation for an organization to become an effective service provider by institutionalizing selected Project and Work Management, Support, and Service Establishment and Delivery processes. Work groups define a service strategy, create work plans, and monitor and control the work to ensure the service is delivered as planned. The service provider establishes agreements with customers and develops and manages customer and contractual requirements. Configuration management and process and product quality assurance are institutionalized, and the service provider also develops the capability to measure and analyze process performance.

10.6.2.3   Maturity Level 3: Defined

At maturity level 3, service providers use defined processes for managing work. They embed tenets of project and work management and services best practices, such as service continuity and incident resolution and prevention, into the standard process set. The service provider verifies that selected work products meet their requirements and validates services to ensure they meet the needs of the customer and end user. These processes are well characterized and understood and are described in standards, procedures, tools, and methods.

10.6.2.4   Maturity Level 4: Quantitatively Managed

At maturity level 4, service providers establish quantitative objectives for quality and process performance and use them as criteria in managing processes. Quantitative objectives are based on the needs of the customer, end users, organization, and process implementers. Quality and process performance is understood in statistical terms and is managed throughout the life of processes.

10.6.2.5   Maturity Level 5: Optimizing

At maturity level 5, an organization continually improves its processes based on a quantitative understanding of its business objectives and performance needs. The organization uses a quantitative approach to understand the variation inherent in the process and the causes of process outcomes. Maturity level 5 focuses on continually improving process performance through incremental and innovative process and technological improvements. The organization's quality and process performance objectives are established, continually revised to reflect changing business objectives and organizational performance, and used as criteria in managing process improvement.

A critical distinction between maturity levels 4 and 5 is the focus on managing and improving organizational performance. At maturity level 4, the organization and work groups focus on understanding and controlling performance at the sub process level and using the results to manage projects. At maturity level 5, the organization is concerned with overall organizational performance using data collected from multiple work groups. Analysis of the data identifies shortfalls or gaps in performance. These gaps are used to drive organizational process improvement that generates measurable improvement in performance.

10.6.3   CMMI—Services

10.6.3.1   Service Agreement

A *service agreement* is the foundation of the joint understanding between a service provider and customer of what to expect from their mutual relationship. The glossary defines a "service agreement" as a binding, written record of a promised exchange of value between a service provider and a customer.

10.6.3.2   Service Request

Even given a service agreement, customers and end users must be able to notify the service provider of their needs for specific instances of service delivery. In the CMMI-SVC model, these notifications are called "service requests," and they can be communicated in every conceivable way, including face-to-face encounters, phone calls, all varieties of written media, and even non-verbal signals.

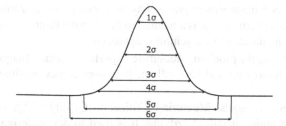

μ ± 1σ = 68.268949213708 % of all values are within the curve
μ ± 2σ = 95.449973610364 % used in demographical statistics
μ ± 3σ = 99.730020393674 %
μ ± 4σ = 99.993665751633 %
μ ± 5σ = 99.999942669685 %
μ ± 6σ = 99.999999802682 % 3.4 defects in 1 million parts

Fig. 10.10 Six Sigma

10.6.3.3 Service Incident

Even with the best planning, monitoring, and delivery of services, unintended events can occur that are unwanted. Some instances of service delivery can have lower than expected or lower than acceptable degrees of performance or quality, or can be completely unsuccessful. The CMMI-SVC model refers to these difficulties as "service incidents." The glossary defines a "service incident" as an indication of an actual or potential interference with a service. The single word "incident" is used in place of "service incident" when the context makes the meaning clear.

CMMI continual improvements and *CMMI services (service request, service incident)* are overlapping standards with ITIL.

10.7 Six Sigma

Six Sigma is a systematic approach for process improvements using analytical and statistical methods. It is assumed that each (business) process can be described as a mathematical function. Six Sigma is a statistical method based on the *normal distribution* (Gauss curve, bell curve). It is a probability distribution of the mean value μ in a large set of measures with the standard deviation σ (tqm 2014) (Fig. 10.10).

The width of the bell curve is determined by the standard deviation of σ. 6 σ simply means that at six times the standard deviation from the mean value 99.9999998026825% of all values are within the normal distribution. For quality assurance in production, this means that with a quality degree of 6 σ, this means that there are only 3.4 defects per 1 million manufactured parts. This is the principle of *zero defects*.

Six Sigma is a management philosophy to transform organizations to become more effective and efficient. It is a method to develop the future business leaders of an organization and uses two disciplined approaches:

DMAIC Defining the problem, Measuring important aspects, Analyzing the information flow, Improving and Controlling. It is used to develop/improve existing practice.

DMADV Defining goals, Measuring Critical to Quality CTQs characteristics, Analyzing, designing details, Verifying. It is used to devise/design a defect-free procedure.

Six Sigma uses varied numerical and problem solving techniques and tools. It is being run by a trained and certified professional (Black Belt) and guided by a Master Black Belt.

10.8   Lean IT

Lean IT is the extension of lean manufacturing and lean services principles to the development and management of information technology (IT) products and services. Its central concern, applied in the context of IT, is the elimination of waste, where waste is work that adds no value to a product or service (Wikipedia 2014).

There are five key Lean IT principles (http://techexcel.com):

Identify Customer and Specify Value
Only a small fraction of the total time and effort in any organization adds value for the end customer. By clearly defining the values for specific products and/or services from the customer's perspective, all the waste can be eliminated.

Identify and Map the Value Stream
A value stream is the activities across all areas of an organization involved in delivering a product or service. This represents the end-to-end process that delivers the value to the customer. Once you have set out the customer requirement, the next step is to identify how you are delivering it to them.

Create Flow by Eliminating Waste
When you are mapping the value stream you will find that only 5–50% of activities add value. Eliminating this waste ensures that your product or service "flows" to the customer without any interruptions, detours or delays.

Respond to Customer Pull
Pull is about understanding customer demand on your services and then tailoring your process to respond to this. Essentially this means you produce only what the customer wants, when the customer wants it.

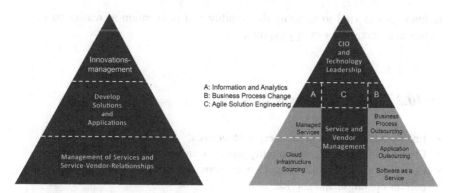

Fig. 10.11 Lean IT (McDonald 2010)

Pursue Perfection

By creating flow and pull that link together, you will find more and more layers of waste become visible. This process continues towards perfection, where every asset and every action adds value for the end customer (Fig. 10.11).

10.9 Conclusions

ITIL, CMMI and Lean IT are suitable methods to support organizational change management for an ERP template rollout.

It takes at least CMMI maturity level 2 to achieve a sufficient organizational readiness. This can be increased during the rollout as part of the project.

ERP implementations have mainly organizational than technological challenges.

10.10 M&A Merger & Acquisition IT Integration

10.10.1 Motivation

The highest volume of activity over the longest period in an integration, particularly large-scale integrations, occurs in the IT environment.

The volume of activity alone increases complexity. Information Technology commonly has the highest number of dependencies on other functions to execute its plans.

An organization's IT integration strategy must be closely aligned with the company's strategic objectives and goals, and further refined to meet the unique needs of each individual business unit.

Building staff commitment to new goals and ways of doing business, and supporting these initiatives through a smooth integration of information technologies is vital to securing the stability and momentum to realize cost efficiency and maximize synergy capture.

10.10.2   M&A Objectives

- Enable business to achieve merger objectives.
- Create standard processes to unify business units and functions from M&A.
- Consolidate IT organizations, IS operations and platforms without interrupting current operation and service.
- Become a source of cost savings through realized synergies.
- Achieve a more aligned information system planning and generate more benefits from IT.
- IT integration effort must essentially aim at

  - leveraging the competencies of both companies,
  - integrating or consolidating business processes and assets
  - eliminating expensive redundancies of data centers and processes.

10.10.3   M&A Challenges

The IT integration plan can either make or break the M&A process.
  Key challenges faced are to

- identify and resolve IT conflicts between organizations,
- assess, analyze and plan the integration of two different IT infrastructures without any operational loss or efficiency,
- improve operational efficiency by identifying the synergies to reduce the *TCO (total cost of ownership)*,
- identify all the touch points of information flow and the data source required for integration,
- maintain the corporate security policies to protect the data and comply with all regulations,
- technology standardization and output quality in disparate IT system scenario.

10.10.4   M&A IT Integration Strategies

To catch the low-hanging saving fruits of M&A, CIOs need to focus on technology consolidation (consolidating data centers, rationalizing vendor contracts, and renegotiating software licenses for a larger base).

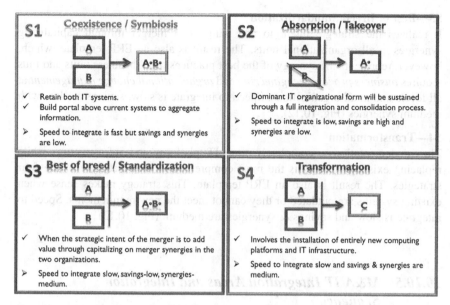

Fig. 10.12 M&A IT integration strategies

Application rationalization is important to eliminate the cost of maintaining redundant and supporting business process integration through a single set of applications.

Processes are the key to acting and looking like a single organization. Processes such as service desk, procurement, security policies, and software development must be stabilized quickly.

In principle, four scenarios can be distinguished:

S1—Coexistence/Symbiosis
Retain both IT systems. Build portals above current systems to aggregate information. The focus is on business process standardization and master data consolidation. This is required for all existing systems separately and needs additional effort. Speed to integrate is fast but savings and synergies are low (Fig. 10.12).

S2—Absorption/Takeover
The dominant IT organizational form will be sustained through a full integration and consolidation process. This is used to develop an ERP template, which is implemented in all business units. All three primary strategic ERP objectives will be achieved. However, this requires a great effort of *business process reengineering* in all business areas in which the new system is implemented, combined with appropriate training and *organizational change management*. Speed to integrate is fast but savings and synergies are low (Fig. 10.12):

S3—Best of Breed/Standardization

It realizes the strategic intent to add value of a merger through capitalizing synergies in all organizational units. The result is also an ERP template, which, however, represents the synergy of the best practices of all business units and thus requires *business process reengineering* and *organizational change management* at all levels and in all organizations. Speed to integrate is slow, savings are low with medium synergies (Fig. 10.12).

S4—Transformation

The implementation a completely new IT platform and infrastructure while replacing existing systems is the most comprehensive and elaborate of the four strategies. The result is also an ERP template. This strategy makes sense when existing systems are outdated or they cannot meet the new requirements. Speed to integrate is slow and savings & synergies are medium (Fig. 10.12).

10.10.5   M&A IT Integration Areas and Integration Sequence

M&A IT integration areas involve business, technology, applications and data (Fig. 10.13).

Security, data, user interface (UI) & applications consolidation are the basis for a successful integration (Fig. 10.14).

Identity Management

- Centralize identity management to ensure security in real time.
- Increases the flexibility and agility of business units for managing employee identities.
- Increases the flexibility and agility of business units for managing identities across company boundaries with customers, distributors, or suppliers.

Data Consolidation

- Aggregates financial data, provides a single view of products, customers and employees.
- Helps to retain customers by focusing on those with the deepest relationships and greatest profit potential.
- Optimizes product portfolio by retaining high-margin products and spinning off the rest.

User Interface Consolidation

- Consolidates user interfaces.

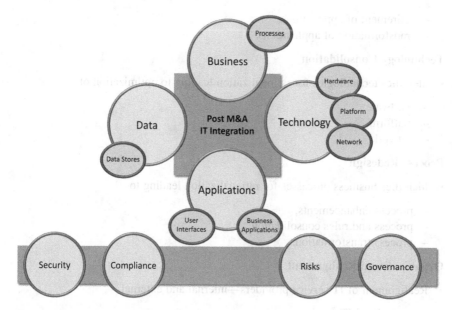

Fig. 10.13 M&A IT integration areas

> Security, Data , UI & Applications Consolidation are the basis for successful integration.

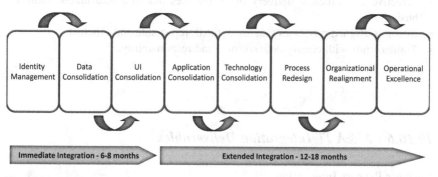

Fig. 10.14 M&A IT integration sequence

- Provides unified view of the business for customers, employees and business partner.

Application Consolidation

- Identifies applications for rationalization leading to
  - retainment of applications from both entities,
  - consolidation of applications,

- retirement of applications,
- transformation of applications.

Technology Consolidation

- Identifies technologies for rationalization leading to optimization of

  - hardware,
  - platform,
  - network.

Process Redesign

- Identifies business processes for rationalization leading to

  - process enhancements,
  - process and rules consolidation,
  - process transformation.

Organizational Realignment

- Realignment of IT service providers—internal and external

  - centralized IT services,
  - vendor consolidation.

Operational Excellence

- Effective and efficient delivery of IT services that add measurable value to business.
- Business aligned processes that are defined, repeatable and efficient.
- Trained staff with clearly defined roles and responsibility.

10.10.6   M&A IT Integration Deliverables

Business Process Integration

- Elimination of process variants and standardization to a global template.

IT Infrastructure Integration

- Integration of data centers, networks, storage, servers.

User Infrastructure Integration

- Integration of E-mail, service desks, enterprise portals, ...

Application Integration

- Consolidation of back-office, middleware and front-office/niche/domain specific applications.

- Functional overlap of applications
  - use business process decomposition to deliver a list of duplicates, overlapping and complementary applications, especially in front-office.
- Technical overlap
  - at middleware and data layers.
- Process overlap
  - at the back-office, ERP layers.

Data Integration

- Archiving of old transaction and master data and migration of active transaction and master data to new schema.

IT Process, Tools Integration

- IT process—2-stage unification plan:
  - IT support process unification,
  - unification of IT development process and advanced support process.

References

Bächle M, Kolb A (2012) Einführung in die Wirtschaftsinformatik. Oldenbourg Wissenschaftsverlag, München

Bucksteeg M (2012) Itil 2011 – der Überblick. Addision-Wesley Verlag, Boston, MA

CMMI (2011) CMMI für Entwicklung, version 1.3. CMMI Product Team, Carnegie Mellon University

CMMI Institute (2014) http://whatis.cmmiinstitute.com. Accessed 6 Jan 2015

Desai S, Srivastava A (2013) ERP to E^2RP a case study approach. PHI Learning Private Limited, Delhi

Heutschi R et al (2004) WebService-Technologien als Enabler des Real-time Business. In: Alt R, Österle H (Hrsg) Real-time business: Lösungen, Bausteine und Potentiale des Business Networking. Springer, Berlin, S. 133–155

ISO/IEC20000 (2012) ISI 20000 White Paper. APMG-International, High Wycombe. http://www.apmg-international.com. Accessed 6 Jan 2015

ITIL Glossary (2011) ITIL Glossar und Abkürzungen Deutsch. https://www.axelos.com/glossaries-of-terms.aspx. Accessed 6 Jan 2015

Johnson C (2005) GLOBE, Nestlé. http://www.nestle.com/assetlibrary/documents/library/presentations/investors_events/investors_seminar_2005/globe_jun2005_johnson.pdf. Accessed 16 Nov 2014

Mathur S (2011) McDonald's spices up products for Indian vegetarians, Budding Markets.com. http://www.buddingmarkets.com/?p=39. Accessed 16 Nov 2014

McDonald M (2010) A model for the lean IT organization. Gartner. http://blogs.gartner.com/mark_mcdonald/2010/06/25/a-model-for-the-lean-it-organization. Accessed 16 Nov 2014

Schemm J (2008) Stammdatenmanagement zwischen Handel und Konsumgüterindustrie -
    Referenzarchitektur für die überbetriebliche Datensynchronisation. Dissertation, Universität
    St. Gallen, Difo-Druck, Bamberg
Schmelzer J, Sesselmann W (2008) Geschäftsprozessmanagement in der Praxis. Carl Hanser
    Verlag, München
Schmidt A (2010) Entwicklung einer Methode zur Stammdatenintegration. Dissertation,
    Universität St. Gallen, Logos Verlag, Berlin
Springer Gabler Verlag [1] (Herausgeber) Gabler Wirtschaftslexikon, Stichwort: ERP, online im
    Internet. http://wirtschaftslexikon.gabler.de/Archiv/3225/erp-v14.html. Accessed 16 Nov 2014
Springer Gabler Verlag [2] (Herausgeber) Gabler Wirtschaftslexikon, Stichwort:
    Geschäftsprozess, online im Internet. http://wirtschaftslexikon.gabler.de/Definition/
    geschaeftsprozess.html. Accessed 16 Nov 2014
tqm (2014) Six Sigma, tqm.com Total quality Management. http://www.tqm.com/beratung/six-
    sigma. Accessed 16 Nov 2014
Watson J (2000) China's big mac attack. Foreign Affair, May/June 2000, 79. Jg.,
    Nr. 3, ABI/INFORM Global S. 120–134. http://www.foreignaffairs.com/articles/56052/
    james-l-watson/chinas-big-mac-attack. Accessed 16 Nov 2014
Wikipedia (2014) Lean IT, Wikipedia.org. http://www.wikipedia.org/wiki/Lean_IT. Accessed
    16 Nov 2014

Chapter 11
SCM: Supply Chain Management

Abstract This chapter describes procedures and methods of *demand forecasting* and *inventory management* as basis for the successful implementation of supply chain management. Demand forecasting focuses on *time series models* and inventory management on *economic order quantity (EOQ)* models. The calculation methods for five time series forecasting procedures and three inventory management models serve as the basis for the automation of the supply chain in part I of this book.

11.1 Definition and Goals

Supply Chain Management (SCM) is the control of material, information, and financial flows within a supply chain from the raw material supplier through the manufacturer, the intermediate trade to the end customer.

Supply chain management systems synchronize the order-to-cash process, i.e. information streams (orders) with goods and services (deliveries) and cash flows (invoices/payments).

The goal of an efficient supply chain management system is to minimize inventories while ensuring delivery.

11.2 Demand Forecasting

Demand forecasting and *inventory management* are two key components for optimizing and automating supply chain management. Demand forecasting is primarily used to suppress the bullwhip effect (Chen et al. 2000). Inventory Management is used to optimize inventory while minimizing inventory costs. Order quantities, order cycle, minimum and maximum inventory levels are determined from this.

Forecasting is a significant element of demand management. It is the basis for the planning and the basis for management decisions. The future planning is a stochastic process and therefore an exact prediction is not possible. The goal is therefore to develop techniques that mimic the gap between the current demand and

© Springer-Verlag GmbH Germany 2017

K.-D. Gronwald, *Integrated Business Information Systems*,

DOI 10.1007/978-3-662-53291-1_11

the forecast. Forecasting is therefore an iterative, repetitive process, which compares the prediction with the reality and adapts accordingly.

Demand Forecasting is the link between Demand Generation (CRM: Sales, Marketing) and Demand Fulfillment (SCM: Manufacturing, Logistics) and is therefore a prerequisite for successful supply chain integration with a periodicity of between one day and one month (depending on industry).

11.2.1 Qualitative Forecasting Methods

Qualitative forecasting methods are based on intuition and experience of the forecaster. They are mainly used for long-term projections, such as the introduction of new products, if current data is not relevant or does not exist. Qualitative methods are for example

- Opinion of an executive committee
- Salesforce experience
- Customer surveys
- Delphi methods

The *delphi method* is a structured iterative communication technique which relies on a panel of experts with the goal to gain as much intelligence as possible from the expert knowledge, without individual opinions dominating. Thus, a group result is achieved, which is based on the consensus of normally non-consensual individual opinions. The procedure requires the following steps (Lenk 2009):

1. **Use of a Questionnaire**
   The actual question is formulated here. The questionnaire is the communication medium between the participants and the moderators.
2. **Interviewing the Experts**
   To answer the questions, the necessary subject matter expertise is assumed. The problem is that experts often disagree.
3. **Anonymity of Expert's Responses**
   On the one hand, the anonymization gives the participants the opportunity to express themselves freely and, on the other hand, it prevents the moderators from intuitively weighting the responses based on the expertise of individual persons.
4. **Determination of the average Group Response**
   The expert responses are evaluated by means of mean-value methods or other examination methods and thus a distribution of the group responses is determined.
5. **Feedback Round(s)**
   The intermediate results will be submitted to the same participants again. Through the confrontation with the group opinion, the participants learn how

their opinions are related to the majority and they will have to take this into consideration for the further argumentation.

6. **Repeated Survey until a Termination Criterion is reached**

The repetition of the survey will result in minimizing the diversity of opinions. Termination criteria can be time dependent or a limited number of iterations or the deviation from the mean value, or any other appropriate criterion.

11.2.2 Quantitative Forecasting Methods

Can be separated into two dominating classes (DecisionCraft 2010) *cause and effect methods* and *time-series methods*.

Cause and effect methods (also referred to as causal methods) are used when one or more external factors (independent variables) are related to the demand and thus influence the forecast, such as income development, economic forecasts, etc. (econometric factors). They sometimes require elaborate statistical methods (regression, econometric models, neural networks). These procedures are covered in the section *Big Data Demand Management*.

Time-series methods assume that the future is a continuation of the past. Historical data are used for predicting the future without considering external factors. In this chapter, we deal exclusively with *time-series methods*. In detail, they are

- Naive forecasting
- Simple moving average forecasting
- Weighted moving average forecasting
- Exponentially smoothing forecasting
- Linear trend forecasting

Common to all these processes is their smoothing effect on fluctuations in demand.

11.2.3 Time Series Forecasting Components

Trend Variations represent changing trends over time. Causes can be population growth, people migrations, income changes, … Trends can be linear, exponential, asymptotic, …. Cyclic variations are wave-like periodical changes that occur over a longer period (>1 year) and are caused by macroeconomic and political factors.

Business cycles such as recession and growth (for example, the euro crisis) or the seasonal distribution of beer consumption are also part of it.

Seasonal fluctuations show periodic highs and lows at specific hours, days, seasons, …

Random fluctuations are caused by unpredictable events such as strikes, natural catastrophes, terrorist attacks, epidemics, . . .

11.2.4   Naive Forecast

Naive forecast assumes that the current demand is equal to the one of the previous period. Since there are no causal connections, this method provides, among other things, inaccurate results. Good candidates, however, are business models whose behavior is determined by random influences (Fig. 11.1).

$$F_{t+1} = A_t$$

with F_{t+1} = forecast for period $t + 1$

A_t = actual demand for period t

11.2.5   Simple Moving Average Forecast

Gives good results for relative stable demands. For n = 1 it is identical with the naive forecast (Fig. 11.2).

$$F_{t+1} = \frac{\sum_{i=t-n+1}^{t} A_i}{n}$$

with F_{t+1} = forecast for the period $t + 1$

n = number of periods, which are used for the calculation of the moving average.

A_i = actual demand in period i

Example: Calculation of the Forecast for Period 5 with a Moving Average over 4 Periods

$$F_5 = \frac{1600 + 2200 + 2000 + 1600}{4} = 1850$$

The actual demand for period 5 is 2500.

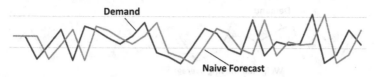

Fig. 11.1 Forecasting—naive forecast

Fig. 11.2 Forecasting—simple moving average

11.2.6   Weighted Moving Average Forecast

Allows for better tracking of changes in the overall picture. The weighting factor is based on the experience of the forecasters. However, although the method responds to changes, it is still a mean value method. Therefore, the process is poorly suited to show trends (Fig. 11.3).

$$F_{t+1} = \sum_{i=t-n+1}^{t} w_i A_i$$

with F_{t+1} = forecast for period $t + 1$

n = number of periods, which are used for the calculation of the moving average.

A_i = actual demand in period i

w_i = weighting factor for period; with $\sum w_i = 1$

Example: calculation of the forecast for period 5 with weighting factors 0.4, 0.3, 0.2, 0.1

$$F_5 = 0.1(1600) + 0.2(2200) + 0.3(2000) + 0.4 * 1600 = 1840$$

The actual demand for period 5 is 2500.

11.2.7   Exponentially Smoothing Forecasting

With this method, the forecast for the demand of the subsequent period is corrected based on the current period by a fraction of the difference between the current demand and the forecast of the subsequent period. This approach requires less data

Fig. 11.3 Forecasting—weighted moving average

than the weighted moving average. Because of its simplicity, this process is one of the most widely used methods. However, like the others, it is not very suitable for seasonal changes as well as for data that show no or only small trends. For $\alpha = 1$ this method is equivalent to the naive forecast.

$$F_{t+1} = F_t + \propto (A_t - F_t)$$

with F_{t+1} = forecast for period $t + 1$

F_t = forecast for period t

A_t = actual demand for period t

$\propto$ = smoothing factor $(0 \leq \propto \leq 1)$

Example 1: calculation of the forecast for period 3.
The forecast for period 2 is 1600. The smoothing factor is $\alpha = 0.3$ (Fig. 11.4).

$$\text{with } F_2 = 1600 \text{ and } \propto = 0.3 \ F_{t+1} \text{ becomes}$$
$$F_{t+1} = F_t + \propto (A_t - F_t)$$
$$F_3 = F_2 + \propto (A_2 - F_2) = 1600 + 0.3(2200 - 1600) = 1780$$

The actual demand for period 3 is 2000.
Example 2: calculating the forecast for period 3.
The forecast for period 2 is 1600. The smoothing factor $\alpha = 0.5$ (Fig. 11.4).

$$\text{With } F_2 = 1600 \text{ and } \propto = 0.3 \ F_{t+1} \text{ becomes}$$
$$F_{t+1} = F_t + \propto (A_t - F_t)$$
$$F_3 = F_2 + \propto (A_2 - F_2) = 1600 + 0.5(2200 - 1600) = 1900$$

The actual demand for period 3 is 2000.

11.2.8 Linear Trend Forecast

Linear trend forecast is a simple linear regression with a trend line based on a series of historical data (Fig. 11.5).

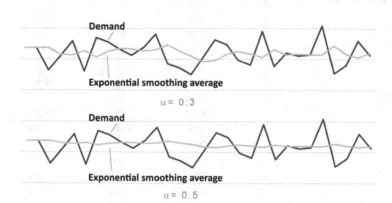

Fig. 11.4 Forecasting—exponential smoothing average

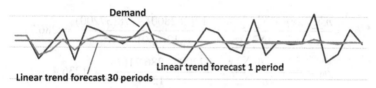

Fig. 11.5 Forecasting—linear trend

$$\widehat{Y} = b_0 + b_1 x$$

with $\widehat{Y}$ = forecast or dependent variable

x = time axis (independent variable)

b_0 = y intercept

b_1 = slope of trend line

$$b_1 = \frac{n \sum (xy) - \sum x \sum y}{n \sum x^2 - \left(\sum x\right)^2}$$

$$b_0 = \frac{\sum y - b_1 \sum x}{n}$$

x = time axis (independent variable)

y = dependent variable

n = number of observations

Example: Demand for Alpha Beer for periods 1–12 is shown in Table 11.1. Looking for the trend line and the forecast for period 13.

Table 11.1 Demand Alpha Beer

Period	Demand	x^2	xy
1	1600	1	1600
2	2200	4	4400
3	2000	9	6000
4	1600	16	6400
5	2500	25	12,500
6	3500	36	21,000
7	3300	49	23,100
8	3200	64	25,600
9	3900	81	35,100
10	4700	100	47,000
11	4300	121	47,300
12	4400	144	52,800
$\sum x = 78$	$\sum y = 37{,}200$	$\sum x^2 = 650$	$\sum xy = 282{,}800$

$$b_1 = \frac{n \sum (xy) - \sum x \sum y}{n \sum x^2 - (\sum x)^2} = \frac{12(282800) - 78(37200)}{12(650) - 78^2} = 286.71$$

$$b_0 = \frac{\sum y - b_1 \sum x}{n} = \frac{37200 - 286.71(78)}{12} = 1236.4$$

The trend line *is* $\widehat{Y} = 1236.4 + 286.7x$

For $x = 13 : \widehat{Y} = 1236.4 + 286.7(13) = 4963.5 = 4964.$

11.3   Inventory Management

11.3.1   Inventory Costs Overview

Inventory costs add to an average of 30–35% of the material value of a product or 6–15% of the revenue depending on the industry (Dietl 2012) (Table. 11.2):
Inventory cost causes are according to Waser (2010)

- **Bound capital**: Purchasing and storage must be pre-financed. For the bound capital interest on current assets must be paid.
- **Infrastructure and handling costs**: Storage infrastructure (including storage and removal) of goods must be made available. Creation, maintenance and depreciation costs.
- **Material management costs**: Material stocks must be regularly checked and re-ordered as required.
- **Depreciation due to aging**: Material in stock loses value. These value adjustments can be significant (date of expiry or innovation).

Table 11.2 Inventory costs of three global companies (examples)

Results 2011	ABB	Novartis	Toyota
Inventory (Mio. $)	5737	5930	15,685
Revenue (Mio. $)	37,990	58,566	228,427
Assets (Mio. $)	39,648	117,496	358,607
Share of total assets (%)	14.5%	5.0%	4.4%
Share of revenue (%)	15.1%	10.1%	6.9%

- **Backorder costs**: Inventories that are too low lead to backorder costs (out-of-stock) in the form of expensive express deliveries or lost sales.
- **Concealed defect costs**: Inventory balances the production processes and thereby obscures potential weak points in production and logistics (kanban, just-in-time).

11.3.2 Inventory Management Models: Overview

According to Dietl (2012) we can distinguish between *single-period models* and *multi-period models*, which can have both *deterministic* and *stochastic* demand structures:

- **Single-period models with deterministic demand** are contractually regulated sales volumes within one period (newspaper subscription).
- **Single-period models with stochastic demand** have an unpredictable demand within one period (newspaper sales in shops).
- **Multi-period models with deterministic demand** are contractual defined sales volumes over several periods (supplier contract).
- **Multi-period models with stochastic demand** have an unpredictable demand over several periods (beer sales in retail shops).

Additionally, we distinguish between

- **Fixed-order quantity model** with fixed (optimal) order quantity depending on the target inventory. The order date is variable when the minimal defined inventory is reached.
- **Fixed-time period model** with variable order quantity, depending on the current consumption. The order time is periodic, at a fixed time interval.

This course uses a *fixed-time period model* with *stochastic demand*.

11.3.3 Economic Order Quantity Models (EOQ): Overview

The method, originally developed by Harris (1913) and Andler (1929) as one formula for determining the *optimal production quantity (lot size),* is now mainly

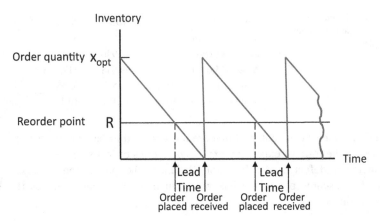

Fig. 11.6 Inventory management EOQ basic model

used as *Economic Order Quantity Model (EOQ)* to determine the *optimal order quantity* while *minimizing inventory costs*. There are a couple of extensions to the basis model. In this course, three models are used, whereby the course participants decide for one of the models based on their results from the first fiscal year (game round 1) and thus must determine the optimal parameters for the supply chain management automation.

The following inventory costs will be taken into consideration here:

- **Carrying costs**—direct costs caused by material storage.
- **Ordering costs**—costs to replenish the inventory.
- **Shortage costs (backorder costs)**—costs if demand cannot be fulfilled due to lacking inventory.

11.3.4 The Basic EOQ: Model

The basic EOQ model is a single formula for determining the optimum order size, which minimizes the sum of inventory costs and order costs (Fig. 11.6).

Assumptions

- Demand is known and relative constant over time.
- No shortages allowed.
- Constant lead time.
- The order quantity is delivered completely.

Determination of the optimal order quantity x_{opt} considering all relevant cost components. Only order costs (B) and inventory costs (L) are considered. The total costs K are

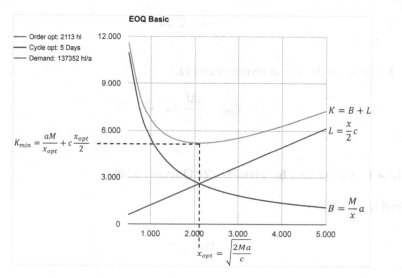

Fig. 11.7 EOQ basic model: calculate minimal inventory cost

$$K = B + L$$

Order costs B are calculated as

$$B = \frac{M}{x} a$$

with M = annual demand; x = order quantity; a = fixed order costs

Inventory costs L are calculated as

$$L = \frac{x}{2} c$$

with $\frac{x}{2}$ = average inventory; c = fixed interests and inventory costs

The optimal order quantity x_{opt} is the minimum of the total costs K (Fig. 11.7). Solution 1: Intersection of L and A and solving for x.

$$\frac{x}{2} c = \frac{M}{x} a \rightarrow x^2 = \frac{2M}{c} a$$

Solution 2: Derivation of the total costs K after the order quantity x.

$$\frac{\partial K}{\partial x} = -\frac{M}{x^2} a + \frac{c}{2} = 0 \rightarrow x^2 = \frac{2M}{c} a$$

$$x_{opt} = \sqrt{\frac{2Ma}{c}}$$

Minimum of the annual inventory costs K:

$$K_{min} = \frac{aM}{x_{opt}} + c\frac{x_{opt}}{2}$$

11.3.4.1 Calculation for Fixed-Order Quantity Model

Total annual orders:

$$A_b = \frac{M}{x_{opt}}$$

Order cycle (days):

$$B_z = \frac{364}{M/x_{opt}} = \frac{364}{A_b}$$

Minimum inventory l_{min}:

$$I_{min} = \frac{x_{opt}}{B_z} * l_z$$

With delivery time l_z in days

Example
Parameters

- Interests and fixed inventory costs c = $ 0.5 per hectoliter
- Fixed order costs a = $20.00 per hectoliter
- Demand M = 84,000 hl

Looking for

- Optimal order quantity x_{opt}
- Minimum inventory
- Annual number of orders
- Order cycle

Optimal order quantity:

$$x_{opt} = \sqrt{\frac{2Ma}{c}} = \sqrt{\frac{2*84,000*20}{0.5}} = 2,592.30 \text{ hl}$$

Annual number of orders (A_b):

$$A_b = \frac{M}{x_{opt}} = \frac{84,000}{2,592.30} = 32.04 \rightarrow 32$$

Order cycle Bz:

$$B_z = \frac{364}{M/x_{opt}} = \frac{364}{A_b} = 11.38 \text{ Tage} \rightarrow 11 \text{ days}$$

Minimal annual inventory costs:

$$K_{min} = a\frac{M}{x_{opt}} + c\frac{x_{opt}}{2} = 20\frac{84,000}{2,592.30} + 0.5\frac{2,592.30}{2} = \$1,296.15$$

The actual inventory cost result in a first approximation by multiplication with the root of the standard deviation of the demand.

Actual inventory costs for an optimized supply chain (Fig. 11.8):
$K_{akt} = K_{min} * \sqrt{462.07} = \$1,296.15 * 21.50 = \$27,867.23$
Actual inventory costs for a non-optimized supply chain (Fig. 11.9):
$K_{Rakt} = K_{Rmin} * \sqrt{2,065.03} = \$1,296.15 * 45.44 = \$58,897.06$

11.3.4.2 Calculation for Fixed-Time Period Model

Example for 30 annual orders:

$$A_b = \frac{M}{x_{opt}} = 30;$$

Order cycle:

$$B_z = \frac{364}{30} = 12.13 \text{ Tage} \rightarrow 12 \text{ days}$$

Optimal order quantity:

$$x_{opt} = \frac{M}{A_b} = \frac{84,000}{30} = 2,800 \text{ hl}$$

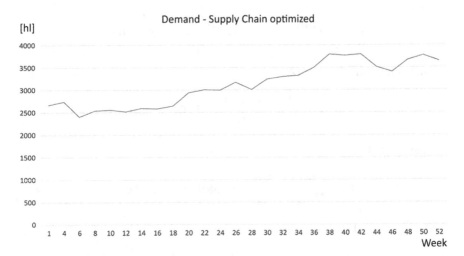

Fig. 11.8 Demand for optimized supply chain

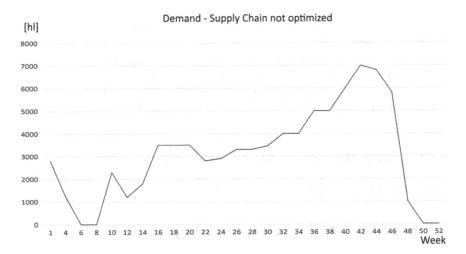

Fig. 11.9 Demand for supply chain not optimized

Minimum inventory:

$$I_{min} = \frac{x_{opt}}{B_z} * l_z = \frac{2,800}{12} * 7 = 1,633 \text{ hl}$$

The optimal order quantity is an approximation, based on cost and demand estimates. It is therefore acceptable to set the value of x to the next higher integer. The EOQ model is robust, since x is a square root and thus errors are attuned when estimating M, c and a.

11.3.5 Non-instantaneous Receipt Model

The order quantity is delivered partially, distributed over the period between two ordering cycles. This results in different values for inventory costs and inventory.

Assumptions

- Demand is known and relative constant over time.
- No shortages allowed.
- Constant lead time.
- The order quantity is delivered partially, distributed over the period between ordering cycles.

The non-instantaneous receipt model is a variation of the basic EOQ model, also known as *gradual usage, production lot size model*. The order quantity is not supplied completely but in several parts distributed over the period between two ordering cycles. This results in different values for storage costs and inventory. This situation is generally found when the storage user is a producer at the same time, or when the reseller and the manufacturer of a product is one and the same (Fig. 11.10).

Special parameters for this model:

p = daily rate of replenished inventory
d = daily rate of depleted inventory

The order cost component of the basic EOQ model remains unchanged. However, the inventory cost component is different since the average stock level is different. The maximum stock level is slightly lower than x (Fig. 11.11).

The maximum inventory MB is:

$$MB = x - \frac{x}{p}d = x\left(1 - \frac{d}{p}\right)$$

The average inventory level ML is half the maximum inventory:

$$ML = \frac{1}{2}MB = \frac{x}{2}\left(1 - \frac{d}{p}\right)$$

Total inventory costs:

$$L = c\frac{x}{2}\left(1 - \frac{d}{p}\right)$$

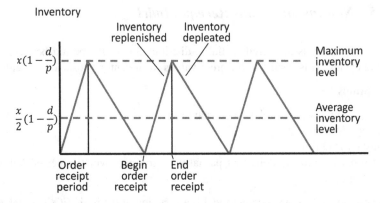

Fig. 11.10 EOQ Inventory management non-instantaneous receipt model

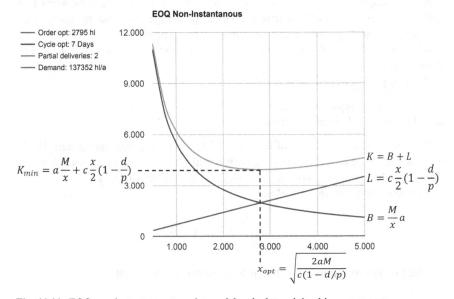

Fig. 11.11 EOQ non-instantaneous receipt model: calculate minimal inventory cost

Total annual costs:

$$K = a\frac{M}{x} + c\frac{x}{2}\left(1 - \frac{d}{p}\right)$$

To calculate x_{opt}, set the inventory costs L equal to the order costs B and solve the equation for x:

$$a\frac{M}{x} = c\frac{x}{2}\left(1 - \frac{d}{p}\right)$$

$$x_{opt} = \sqrt{\frac{2aM}{c(1 - d/p)}}$$

Example

The supply chain has its own factory.

It is assumed that the order costs are equal to the costs to start the production costs.

With c = $0.5 per hectoliter and MR = 84,000 hl per year for the retailer, the daily inventory depletion rate is

$$d = \frac{84,000}{364} = 230.77 \text{ hl}$$

The production rate with which the inventory will be replenished is

$$p = 800 \text{ hl}$$

The optimal order quantity becomes

$$x_{opt} = \sqrt{\frac{2aM}{c(1 - d/p)}} = \sqrt{\frac{2*20*84,000}{0.5(1 - 230.77/800)}} = 2,173 \text{ hl}$$

The total annual minimum inventory costs are

$$K_{min} = a\frac{M}{x} + c\frac{x}{2}\left(1 - \frac{d}{p}\right)$$

$$K_{min} = 20\frac{84,000}{2,173} + 0.5\frac{2,173}{2}\left(1 - \frac{230.77}{800}\right) = \$1,159$$

The period to continually replenish the inventory is

$$B_p = \frac{x_{opt}}{p} = \frac{2,173}{800} = 2.71 \text{ days} \rightarrow 3 \text{ days}$$

Total number of production cycles (order cycles):

$$P_l = \frac{M}{x_{opt}} = \frac{84,000}{2,173} = 38.66 \rightarrow 39$$

Maximum inventory M_l:

$$M_l = x_{opt}\left(1 - \frac{d}{p}\right) = 2,173\left(1 - \frac{230.77}{800}\right) = 1,546 \text{ hl}$$

11.3.6   Shortages Model

The basic EOQ model does not allow shortages. The shortages model does allow this explicitly. However, it is assumed that the total demand will be delivered including the shortages as backorder (Fig. 11.12).

Assumptions

- Demand is known and relative constant over time.
- Shortages allowed.
- Constant lead time.
- The order quantity is delivered completely.

Since backordered demand or shortages (S) are balanced as soon as the inventory is filled, the maximum inventory never reaches X, but X-S. Thus, the cost of shortages is inversely proportional to the storage costs. As the order quantity X grows, the inventory costs increase and the shortages costs decrease correspondingly (Fig. 11.13).

The respective cost functions are as follows, with the backorder (shortage) S and the annual backorder costs f per unit:

$$\text{Total backorder (shortage) costs} : F = f\frac{S^2}{2x}$$

$$\text{Total inventory costs} : L = c\frac{(x - S)^2}{2x}$$

$$\text{Total order costs} : B = a\frac{M}{x}$$

$$\text{Total costs} : K = f\frac{S^2}{2x} + c\frac{(x - S)^2}{2x} + a\frac{M}{x}$$

The three cost components do not intersect at the same point. Therefore, the only way to determine the optimal order quantity and the optimal backorder is by differentiating the total-cost curve to x and S.

Solving both equations to x and S respectively will give the results for the optimal order quantity x_{opt} and the optimal shortage S_{opt}:

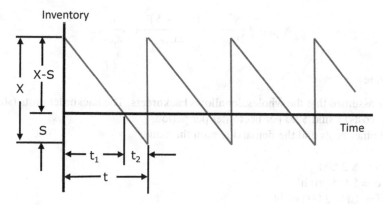

Fig. 11.12 Inventory management: EOQ shortages model

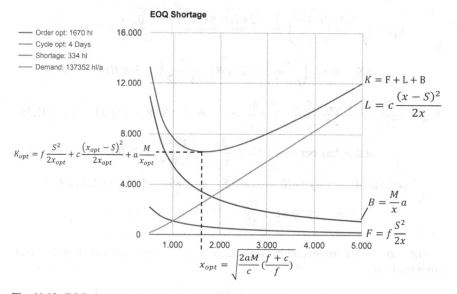

Fig. 11.13 EOQ shortages model: calculate minimum inventory cost

$$x_{opt} = \sqrt{\frac{2aM}{c}\left(\frac{f+c}{f}\right)}$$

$$S_{opt} = x_{opt}\left(\frac{c}{c+f}\right)$$

$$K_{opt} = f\frac{S^2}{2x_{opt}} + c\frac{(x_{opt} - S)^2}{2x_{opt}} + a\frac{M}{x_{opt}}$$

Example

- It is assumed that the wholesaler allows backorders. The backorder costs (shortage costs) f sind \$2.0 per hectoliter per period.
- All other costs and the demand remain the same:

  a = \$ 20.00
  c = \$ 0.5 pro hl
  f = CHF 2.00 pro hl
  M = 84,000 hl

$$x_{opt} = \sqrt{\frac{2aM}{c}\left(\frac{f + c}{f}\right)} = \sqrt{\frac{2*20*84,000}{0.5}\left(\frac{2 + 0.5}{2}\right)} = 2,898 \text{ hl}$$

$$S_{opt} = x_{opt}\left(\frac{c}{c + f}\right) = 2,898\left(\frac{0.5}{0.5 + 2}\right) = 579.6 \text{ hl}$$

$$K_{opt} = f\frac{S_{opt}^2}{2x_{opt}} + c\frac{(x_{opt} - S_{opt})^2}{2x_{opt}} + a\frac{M}{x_{opt}} = 57.96 + 463.68 + 579.71 = \$ 1,101.35$$

$$\text{Total orders per year} : M_b = \frac{M}{x_{opt}} = \frac{84,000}{2,898} = 28.99 \rightarrow 29$$

$$\text{Maximum inventory} : ML = x_{opt} - S_{opt} = 2,898 - 579.6 = 2,318.4 \text{ hl}$$

$$\text{Order cycle} : t = \frac{364}{M_b} = 12.55 \text{ days} \rightarrow 13 \text{ days}$$

The period t_1 where inventory is available and the shortage period t_2 within one order cycle are:

$$t_1 = \frac{x_{opt} - S_{opt}}{M} = 364*\frac{2,898 - 579.6}{84,000} = 10.04 \text{ days} \rightarrow 10 \text{ days}$$

$$t_2 = \frac{S_{opt}}{M} = 364*\frac{579.6}{84,000} = 2.512 \text{ days} \rightarrow 3 \text{ days}$$

References

Andler K (1929) Rationalisierung der Fabrikation und optimale Losgrösse. R. Oldenbourg, München

Chen F et al (2000) The impact of exponential smoothing forecasts on the bullwhip effect. Naval Res Logist 47(4):269–286

DecisionCraft (2010) Choosing the right forecasting technique. DecisionCraft. http://www.decisioncraft.com/dmdirect/forecastingtechnique.htm. Accessed 16 Nov 2014

Dietl H (2012) Operations management. Universität Zürich. http://www.business.uzh.ch/professor ships/som/stu/Teaching/F2012/BA/BWL/5_Lagerhaltungsmanagement.pdf. Accessed 16 Nov 2014

Harris F (1913) How many parts to make at once factory. Mag Manage 10(2):135–136, 152

Lenk T (2009) Die Delphi-Method in der Regionalentwicklung, Arbeitspapier Nr. 41. Universität Leipzig, Wirtschaftswissenschaftliche Fakultät

Waser B (2010) Hochschule Luzern

Chapter 12
CRM: Customer Relationship Management

Abstract Customer Relationship Management was introduced in Part I. This chapter contains the detailed description of the individual methods, additional remarks on the topics of the CRM strategy and customer relationship, as well as a detailed discussion about the failure of many CRM concepts. The failure rate for CRM implementations is significantly higher compared to other IT projects.

12.1 CRM Strategy

Customer Relationship Management is the establishment of customer relationships with the primary objectives of customer acquisition, the expansion of an existing customer base, the identification of profitable customers and customer retention (loyalty management) (Schmid and Bach 2000).

The aim of a CRM strategy is to build up as much knowledge as possible about customers to use this knowledge to optimize the interaction between companies and customers, with a view to maximizing the *customer lifetime value* (CLV) for the enterprise. By focusing on the economic value of individual customers, there must be a synergy between the company's offerings and the needs, behavior and characteristics of a customer. Customer relationship is an interactive dialogue. Customer satisfaction is a critical parameter. This makes CRM a continuous process for a growing customer centric focus of a company (Kumar and Reinartz 2012).

12.2 CRM: Customer Retention

There have been recent discussions about whether customer loyalty is still useful or not. The studies and opinions differ widely, with the cost factor being put into the foreground. Customer relationship management has not just financial aspects, although the result is reflected in profitability. It seems that companies are setting different standards here than customers. For companies, the best customer seems to be the most profitable one.

It is undisputed that it is cheaper to retain an existing customer than to sell to new ones. The list of reasons for customer loyalty therefore does not include any figures

K.-D. Gronwald, *Integrated Business Information Systems*,
DOI 10.1007/978-3-662-53291-1_12

but trends. In 2010, the Chartered Institute of Marketing did a survey on the cost of customer acquisition versus the cost of customer retention. The data showed that it is 4–30 times more expensive to acquire new customers than to retain existing ones (The Chartered Institute of Marketing 2010).

There are a couple of arguments for building a customer-centric marketing strategy:

- It is significantly more expensive to sell to a new customer than to an existing one.
- A typical unsatisfied customer will affect a large number of other people.
- Most complaining customers will remain when their problem has been resolved.

12.2.1   Strategic CRM

The goal of *Strategic CRM* is to build as much knowledge as possible about customers, to use this knowledge to optimize the interaction between companies and customers, with the aim of maximizing the *Customer Lifetime Value (CLV)* for the company (Kumar and Reinartz 2012).

12.2.2   Analytical CRM

Analytical CRM uses customer data to form profitable relationships between customers and companies. It uses traditional business intelligence (BI) methods such as data warehouse, data mining, and online analytical processing systems (OLAP) to determine customer satisfaction and active measures to optimize the corresponding parameters. In this course, the customer loyalty plays a particular role. This can be measured directly through the *share of wallet*. It means how much percent of its beer consumption a customer shares between the four brands. Marketing measures can be derived from this (Operative CRM).

12.2.3   Loyalty Management and Share of Wallet

In this course, the customer loyalty plays a significant role. This is directly measurable via the *share of wallet*. It means how much percent of its beer consumption a customer shares between the four brands. Marketing measures can be derived from this and are transferred to the operational CRM. This is accompanied by a paradigm shift in marketing from *transactional marketing* to *relationship marketing*, from an *increase in market share* to *share of wallet* (Trommsdorf 2011) (Fig. 12.1).

Fig. 12.1 Paradigm shift in marketing (Ahlert et al. 2000)

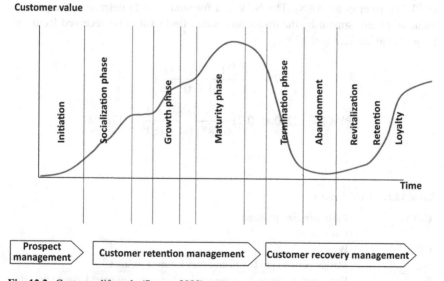

Fig. 12.2 Customer lifecycle (Strauss 2000)

The *share of wallet* is the share of a customer's purchasing power that remains with the company. It is measured by the *customer lifetime value* (CLV), which covers the entire lifecycle of a business relationship (Trommsdorf 2011) (Fig. 12.2).

12.2.4   Customer Lifetime Value CLTV

Customer lifetime value (CLTV) (Wisner et al. 2008) (Tables 12.1 and 12.2).

$$CLTV = \sum_{t=1}^{T} \frac{d_{kt}y_{kt} - F_{kt}}{(1+r)^{t-1}}$$

$$CLTVK1 = \sum_{t=1}^{5} \frac{0.22*22,000 - 1}{(1+0.08)^{t-1}} = 20,866$$

$$CLTVK2 = \sum_{t=1}^{15} \frac{0.15*16,000 - 1}{(1+0.08)^{t-1}} = 27,152$$

12.2.5   Customer Lifetime Value NPV

The customer lifetime value can alternatively be calculated by the *net present value (NPV)* (Wisner et al. 2008). The NPV is a formula used to determine the present value of an investment by the discounted sum of all cash flows received from the project (Tables 12.3 and 12.4).

$$NPV = a \left[\frac{(1+i)^n - 1}{i(1+i)^n} \right]$$

$$NPVK1 = (22,000*0.2) \frac{(1+0.08)^5 - 1}{0.08*(1+0.08)^5} = 17,568$$

Table 12.1 CLTV parameter

CLTV:	Customer lifetime value
k:	Customer
t:	Period
d_{kt}:	Margin per sold product to customer k during period t
F_{kt}:	Fixed costs resulting from the relationship to the customer k during period t
r:	Interests
y_{kt}:	Number of products sold to the customer k during period t

Table 12.2 CLTV customer comparison

Revenue year y	Margin d	Lifetime T
K1: $22,000	20%	5 years
K2: $16,000	15%	15 years

Table 12.3 NPV parameter

a:	Average annual margin
i:	Annual discount rate (8%)
n:	Expected customer lifetime in years

Table 12.4 NPV customer comparison

Revenue/Jahr y	Margin d	Lifetime T
K1: $22,000	20%	5 years
K2: $16,000	15%	15 years

$$NPVK2 = (16,000*0.15)\frac{(1+0.08)^{15} - 1}{0.08*(1+0.08)^{15}} = 20,543$$

12.3 Operative CRM

Operative CRM implements the identified measures of strategic CRM which were quantified in analytical CRM in the form of (automated) solutions for marketing, sales and services. This course deals exclusively with *campaign management* as a basis for appropriate marketing campaigns. It includes

- front office,
- back office,
- basic functions for sales, marketing, service,
- interfaces to other systems (for example to ERP),
- central customer database,
- contact and complaints management,
- reporting.

Front office includes functionalities that support all communication activities with customers, such as:

- information exchange on new products,
- salesforce support.

Back office integrates CRM and SCM functions like

- invoicing
- order processing,
- delivery.

Salesforce management is the sales oriented part of the operative CRM and includes functions

- proposal management,
- account manager management,

- sales regions management,
- opportunity management,
- tele sales management,
- contract management.

After Sales covers loyalty management functions and includes

- triggers for support, delivery and invoice processing,
- support for maintenance and repairs,
- support for complaint management,
- management of key performance indicators (KPI) for customer satisfaction and service level management (SLA).

Marketing includes

- customer segmentation,
- campaign Management,
- lead management,
- product portfolio management.

Customer segmentation is the process of identifying and classifying individual customers and customer groups for targeted marketing and sales activities. This includes

- sociodemographic data,
- data on customer behavior,
- data on the customer value,
- psychological data,
- geographical data.

From a CRM perspective, the customer segmentation is focusing on marketing activities with the aim to optimizing the return on investment (ROI) of sales activities.

Campaign management has the goal to optimize the profitability of marketing activities, reduce marketing costs per customer and increase customer value. Actions are

- specification of product offers to customers,
- definition of the communication channels for the product offering,
- time windows in which the campaigns take place,
- recording of potential customer reactions,
- determine how these reactions are registered in information systems,
- follow-up activities in response to customer responses.

12.4 Communicative CRM

Communicative CRM includes the management of all communication channels between the customer and the company (telephony, internet, e-mail, direct mailing, etc.). The various communication channels are synchronized, controlled and targeted to enable bidirectional communication between customers and companies. This approach is also referred to as *multichannel management* (Grabner-Kräuter and Schwarz-Musch 2009, p. 184).

12.5 Why CRM Projects Fail

The failure rate for IT projects is traditionally high and has been almost constant for many years, including the reasons for its failure. The Standish Group has been publishing its annual Chaos Manifesto for more than 20 years, a record of success and failure of IT projects (Johnson et al. 2013). On average, only 34% of all software projects between 2004 and 2012 were successfully completed (on time, on budget, on target). 46% had to be improved due to budget or time overrun, as well as not achieving project goals, and 20% of all projects failed completely. The reasons for the failure have constantly been the same over the years:

1. The absence of executive management support
2. Insufficient involvement of end users
3. Lack of optimization of project objectives, time, effort and cost estimates, expectation, and so on.

CRM projects have a disproportionally high failure rate of around 50% (Krigsman 2009). In 2013, it was even 63% (Prezant 2013). The main reasons for failure differ also significantly from those for other IT projects (techtarget.com 2004):

1. Lack of cross-departmental and cross-functional coordination
2. No CRM business strategy
3. Missing process changes
4. Lack of senior executive support

The combination of CRM and SCM, demand generation and demand fulfillment (SCM) play a prominent role in the process simulations in this book.

12.5.1   Case Study: CRM Contributes to a Scary Halloween for Hershey

This case study is a good example for the economic impact if demand generation and demand fulfilment don't match. Root cause is the failure of a CRM initiative with a combination of project management issues and wrong executive decisions.

Candy producers record 40% of their annual sales between October and December. Halloween, the biggest candy-consuming holiday, accounts for about $2 billion in sales. For a candy producer, missing Halloween is like a toy company missing Christmas. Unfortunately, in 1999, that's just what happened to Hershey, the nation's largest candy maker. Just before the big candy season, shelves at warehouses and retailers lay empty of treats such as Hershey bars, Reese's Peanut Butter Cups, Kisses, Kit-Kats, and Rolos. Though inventory was plentiful, orders had not arrived and distributors could not fully supply their retailers.

Hershey announced in September that it would miss its third quarter earnings forecasts due to problems with new customer order and delivery systems that had been recently rolled out. The new enterprise resource planning (ERP) and CRM processes and technology implemented earlier in the year had affected Hershey's ability to take orders and deliver product. The $112 million system aimed to modernize business practices and provide front-to-back automation from order-taking to truck-loading, but Hershey lost market share as problems allowed rivals to benefit during the season. Mars and Nestlé both reported unusual spurts of late orders as the Halloween season grew nearer. The most frustrating aspect of the situation is that Hershey had plenty of candy on hand to fill all its orders. It just couldn't deliver the orders to customers.

By December 1999, the company announced it would miss already lowered earnings targets. It stated that lower demand in the last few months of the year was in part a consequence of the earlier fulfillment and service issues. Hershey had embarked on the project in 1996 to better coordinate deliveries with its retailers, allowing it to keep its inventory costs under control. The company also needed to address Y2K problems with its legacy systems. CRM, ERP, and supply chain management systems were implemented, along with 5000 personal computers and a complex network of servers. The intention was to integrate these software and hardware components in order to let the 1200-person sales force shepherd orders step-by-step through the distribution process. Sales could also better coordinate with other departments to handle every issue from order placement to final delivery. The system was also designed to help Hershey measure promotional campaigns and set prices, plus help run the company's accounting operations, track ingredients, and schedule production and truck loading.

Hershey realized that the business process changes involved with such a transformation were highly intricate. However, despite the size and complexity of the undertaking, the firm decided on an aggressive implementation plan that entailed a large piece of the new infrastructure going live at the same time. Unfortunately, the project ran behind schedule and wasn't ready until July 1999 when the Halloween orders had already begun to come in. Problems in getting customer orders into the system and transmitting the correct details of those orders to warehouses for shipping began immediately. By August, the company was 15 days behind in filling orders, and in September, order turnaround time was twice as long as usual.

In recent years, Hershey sales growth had exceeded its rivals, and the company was expecting 4–6% growth that year. However, sales instead slipped and the company admitted that problems with the new system alone had reduced sales by $100 million during the period.

12.5.2   Impact Factors of Failed CRM Projects on Company Performance (techtarget.com 2004)

Financial Performance

- Market share and operating losses
- Failure to achieve a return on investments
- Budget overruns
- High post-implementation running costs

Customer Service Quality

- Customer confusion, frustration, and dissatisfaction
- Lower service levels
- Slower time to market
- Negative brand perception

Sales Effectiveness

- Lower sales force productivity
- Increased sales force cynicism toward new systems
- Increased sales force turnover

Cultural Impacts

- Low morale within IT and affected departments
- Growing cultural cynicism within the company toward adopting business change
- Company-wide loss of confidence in its ability to enact change
- Lost jobs in the executive suite
- Propensity for companies to become overly conservative with regard to investments in strategic initiatives. This leads to dampened innovation, a failure to strengthen advantages, and deferring the update of aging processes and infrastructure

References

Ahlert D et al (2000) Markenmanagement im Handel. Gabler, Wiesbaden

Grabner-Kräuter S, Schwarz-Musch A (2009) CRM Grundlagen und Erfolgsfaktoren. In: Hinterhuber H, Matzler K (Hrsg) Kundenorientierte Unternehmensführung, Kundenorientierung – Kundenzufriedenheit – Kundenbindung, 6. Auflage. Gabler, Wiesbaden, pp 174–189

Johnson J (2013) The CHAOS Manifesto. The Standish Group. http://www.versionone.com/assets/img/files/CHAOSManifesto2013.pdf. Accessed 5 Jun 2014

Krigsman M (2009) CRM failure rates: 2001–2009. zdnet.com. http://www.zdnet.com/article/crm-failure-rates-2001-2009. Accessed 20 Feb 2015

Kumar V, Reinartz W (2012) Customer relationship management concept, strategy, and tools. Springer, Berlin

Prezant J (2013) 63% of CRM initiatives fail, direct marketing news. Haymarket Media. http://www.dmnews.com/63-of-crm-initiatives-fail/article/303470. Accessed 20 Feb 2015

Schmid E, Bach V (2000) Customer relationship bei Banken, Bericht Nr. BE HSG/CC BKM/4. Universität St. Gallen

Strauss B (2000) Perspektivwechsel: Vom Produkt-Lebenszyklus zum Kundenbeziehungs-Lebenszyklus. In: Thexis 2/2000, p 15–18

techtarget.com (2004) A review of CRM failures. http://media.techtarget.com/searchCRM/downloads/CRMUnpluggedch2.pdf. Accessed 20 Feb 2015

The Chartered Institute of Marketing (2010) Cost of customer acquisition vs customer retention. http://www.camfoundation.com/PDF/Cost-of-customer-acquisition-vs-customer-retention.pdf. Accessed 16 Nov 2014

Trommsdorf V (2011) VL Strategisches Marketing – Markteintritt und Kundenbindung. https://www.marketing.tu-berlin.de/fileadmin/fg44/download_strat/ws1112/08_Markteintritt_und_Kundenbindung.pdf. Accessed 15 Mar 2015

Wisner D et al (2008) Principles of supply chain management. South Western

Chapter 13
BI: Business Intelligence

Abstract Business Intelligence will be formally introduced as an independent discipline with the procedures that distinguish this area from Big Data Analytics: OLAP, OLTP, ETL and Data Mining. Data Mining will be explained using case studies.

13.1 Introduction and Definitions

Business Intelligence (BI) is an umbrella term for applications, infrastructure, tools, and best practices for accessing and analyzing data and information for (strategic) decision-making and performance improvement to achieve market advantages (Gartner 2013). Business intelligence systems belong to the management support systems.

Management support systems (MSS) or management support systems are all IT application systems that support the management, i.e. the specialist and decision makers of a company, in its various tasks. In doing so, these are mainly activities that are used to plan, organize, manage and control operational performance processes. Classical management systems are MIS (management information systems), DSS (decision support systems), EIS (executive information systems) (Gluchowski et al. 2008). The transition to business intelligence (BI) is fluid (Enzyklopädie der Wirtschaftsinformatik 2013).

The quality of the activities of the specialists and managers of a company is decisively determined by the appropriate assessment of current and future external and internal factors as well as by the ability to derive early success relevant decisions for their own company. The more managers know about the available action alternatives and their impact on the underlying target system, the better their decision will be. Data, information, knowledge and communication play an important role in carrying out technical and management tasks (Gluchowski et al. 2008). Information management is the sum of all management tasks in an organization based on its computer based information and communication system (Gluchowski et al. 2008; Gabriel and Beier 2003). Business intelligence systems have evolved evolutionarily from decision support systems (Desai and Srivastava 2013).

© Springer-Verlag GmbH Germany 2017
K.-D. Gronwald, *Integrated Business Information Systems*,
DOI 10.1007/978-3-662-53291-1_13

Definition Business intelligence (BI) is a process that creates knowledge about its own and foreign positions, potentials and perspectives from fragmented, inhomogeneous company, market and competition data (Grothe and Gentsch 2000).

Gluchowski et al. (2008) have formulated the process of *knowledge discovery*) in a phase wise and potentially recursive process as follows:

1. **Selection**
   The selection of the data source to be examined is determined by the objective of knowledge discovery.
2. **Preparation**
   The data is modified so that it is accessible to a subsequent analysis.
3. **Analysis**
   In the analysis, potentially interesting patterns of relationships (regularities, abnormalities) are distilled from the data and described by logical and/or functional dependencies. This phase is called *data mining*.

Definition *Data mining* is an interdisciplinary research approach that finds its roots in statistics, mathematics and artificial intelligence (Gluchowski et al. 2008).

13.2   OLAP and OLTP

The traditional business intelligence approach uses *data mining* systems (*business warehouse*), data transformation (*ETL*), and analytical processing systems (*OLAP*) (Fig. 13.1).

OLTP (Online Transaction Processing) are transaction oriented systems that process the company's operational data with (standardized) software solutions and store them in *normalized* databases or flat files.

ETL (Extract, Transform, Load) is the process responsible for pulling data out of source systems (OLTP) and placing it into a business warehouse (data warehouse).

Business Warehouse contains the data prepared for analytics, reporting and data mining from the ETL process. These data are multidimensional and de-normalized.

OLAP (Online Analytic Processing) accesses multi-dimensional or relational data from the business warehouse for analysis and data mining.

OLAP Data Cube is a core component of the OLAP system. The OLAP cube aggregates facts from each level of a dimension. Dimensions are e.g. products, time, regions, turnover, profit, ... OLAP cubes can have any number of dimensions according to which the business warehouse can be evaluated. In this respect, the term *cube* is not quite correct and misleading.

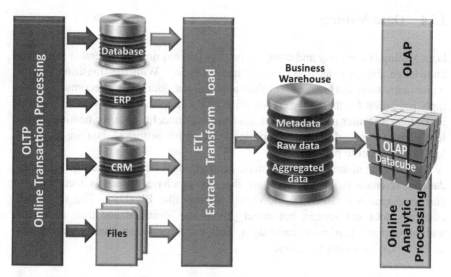

Fig. 13.1 Business intelligence OLTP–ETL–OLAP process flow

13.3   ETL Process

The ETL process is a four-step process that processes the data for the business warehouse. Originally IT-centric, this process is today increasingly carried out by data scientists on the business side, since the process can already be a pre-interpretation of the data before they enter the analysis process (Bächle and Kolb 2012):

1. Cleaning

   - unification of data formats
   - correction of syntactic and semantic deficiencies

2. Harmonizing

   - eliminate synonyms and homonyms as well as different encodings

3. Compacting

   - summation on different aggregation levels that are stored in the business warehouse for performance reasons

4. Enriching

   - calculation and storage of important key figures

With the increasing use of in-memory computing, the separation of transaction-oriented (OLTP) and analysis-oriented (OLAP) queries as well as the complex ETL process are increasingly questioned. Plattner (2013) proposes the combination of OLAP and OLTP, which makes the ETL process obsolete.

13.4   Data Mining

Data Mining is a lengthy and complex process that requires considerable statistical knowledge and an extensive business understanding. With the transition to customer orientation and big data real-time applications, data mining becomes increasingly important for the pattern recognition of the purchasing behavior and for the immediate measures derived from it. Data mining, modeling and verification are the first steps in successful Big Data projects. Extensive historical data are analyzed looking for specific patterns.

Data mining alone without big data, i.e. traditional business intelligence applications, is characterized by the fact that the purchase pattern is identified by customer groups and not by individual persons like Big Data. The individual customer does not appear, but target groups are identified and then addressed with appropriate marketing campaigns. This can have unexpected consequences, as the following example shows.

13.4.1   Case Study: Target Data Mining Figured Out a Teen Girl Was Pregnant Before Her Father Did (Hill 2012)

Every time you go shopping, you share intimate details about your consumption patterns with retailers. And many of those retailers are studying those details to figure out what you like, what you need, and which coupons are most likely to make you happy.

The American retail chain TARGET for example, has figured out how to data-mine its way into your womb, to figure out whether you have a baby on the way long before you need to start buying diapers.

Target assigns every customer a Guest ID number, tied to their credit card, name, or email address that becomes a bucket that stores a history of everything they've bought and any demographic information Target has collected from them or bought from other sources.

Using that, data scientists looked at historical buying data for all the ladies who had signed up for Target baby registries in the past.

They ran test after test, analyzing the data, and before long some useful patterns emerged. Lotions, for example. Lots of people buy lotion, but one data scientist noticed that women on the baby registry were buying larger quantities of unscented lotion around the beginning of their second trimester. Another analyst noted that sometime in the first 20 weeks, pregnant women loaded up on supplements like calcium, magnesium and zinc. Many shoppers purchase soap and cotton balls, but when someone suddenly starts buying lots of scent-free soap and extra-big bags of cotton balls, in addition to hand sanitizers and washcloths, it signals they could be getting close to their delivery date.

As Target's computers crawled through the data, they were able to identify about 25 products that, when analyzed together, allowed them to assign each shopper a "pregnancy prediction" score. More important, they could also estimate her due date to within a small window, so Target could send coupons timed to very specific stages of her pregnancy. So, Target started sending coupons for baby items to customers according to their pregnancy scores.

One day an angry man complained about sending baby coupons to his daughter, who was still in high school, asking if Target wanted to encourage her to get pregnant. The Target manager did not have any idea what the man was talking about, but apologized. A couple of days later, he called to apologize again. On the phone, though, the father was somewhat abashed. He said, that he had talked to his daughter and found out that there had been "some activities" in his house he had not been aware of and that she actually was pregnant and that he had to apologize.

To avoid situations like this, they started mixing in all these ads things they knew pregnant women would never buy, so the baby ads looked random. They put an ad for a lawn mover next to diapers, or a coupon for wineglasses next to infant clothes. That way, it looked like all the products were chosen by chance.

13.4.2   Case Study: Tesco Data Mining for Realtime Inventory Management and Forecasting (Swabey 2013)

The British retail chain Tesco has grown its analytics team to more than 50 scientists. It is staffed mainly by science and engineering graduates, who Tesco trains up in retail expertise and SQL programming skills and who mostly use mathematics application Matlab to conduct their analysis.

The team's biggest wins so far include a statistical model that predicts the impact of the weather on customer buying behavior. By comparing historical weather data with sales records in its 3000 plus stores, Tesco can now adjust its stock levels based on the weather forecast, so its stores do not run out of the goods people want.

There are goods that sell more when the weather is hot, such as barbecue meat, and goods that sell more when is cold, such as cat litter (cats are less likely to leave the house in the winter, Tesco has found). There are other products that sell when it is hold or cold, such as firelighters, but when it is in the middle.

And people are more likely to break out the barbecue when a sunny day follows a prolonged cold spell. By adding this effect to model Tesco reduced out of stock for good weather products by a factor of four.

An even bigger win has come from analyzing discounts and promotions. Tesco runs 1000s of promotions every day, and needs to predict how popular they will be in make it has enough—but not too much—stock to meet demand.

Until recently, it was up to each store's stock controller to estimate how popular each promotion would on a daily basis. But with so many promotions running at any one time, the accuracy of their predictions was understandably limited.

The supply chain analytics team took all the data Tesco had about its historical promotions, and built a detailed predictive model. This pulled in all manner of variables, including the positioning of a discounted product in a store and what other offers were operating at the same time.

That has revealed some interesting insights. For example, a "buy one, get one free" offer works better than a 50% discount for non-perishable goods, such as a cooking sauces, but the revers is true for fruit and vegetables.

It has cut the number of instances of products on promotion being out of stock by 30%.

13.4.3 Case Study: The LAPD Is Predicting Crimes Before They Happen

With the use of computer analytic technology, the Los Angeles Police Department (LAPD) is predicting crimes before they happen. The Real-Time Analysis and Critical Response Division in downtown LA is the hub for this activity. The department has rows of crime analysts and technologists who observe news broadcasts, security camera footage, and maps of the most recent crimes in the city. By recording and mapping the data from previous crimes, an algorithm takes that data and estimates the likelihood of a crime occurring in that area again, and when it is most likely to happen (Loubriel 2014).

The algorithm used by the LAPD and the method is known as predictive policing. The analysis of more than 13 million crimes in 80 years, which are stored in the LAPD data base, were searched for a pattern in a data mining project that allows for a prediction. Criminals are usually very territorial. If they have committed crimes in an area without being caught, it is very likely that they will do it again (Loubriel 2014).

The project has been developed in collaboration with a researcher at the University of California, Los Angeles (UCLA), a professor of anthropology, Jeff Brantingham (Brantingham 2014) whose primary research deals with remote hunt-gatherer tribes in China and a mathematician, George Mohler (2014), whose specialty is the prediction of earthquakes. A wonderful example of the interdisciplinary approach and the necessary abstraction ability in predictive analytics. The competence of the anthropologist to recognize human behavior patterns led to a recognition of patterns in the 13 million records of the LAPD past (Cox 2014).

The next step was the search for a mathematical model that could predict the future from the patterns of the past. Just as little as a great earthquake is predictable, it is also the time and place of a crime that occurs for the first time. In the aftershocks, however, this looks different. They are very predictable and there are mathematical models and algorithms for them (Mohler 2014). The model was tested with the results of the crime analysis by Brantingham and a large agreement was found. The big data step, the real-time application in predictive analytics, followed.

LAPD gets a map every morning with marked squares in the size of 140 × 140 m, in which crimes are most likely predicted. The task is to haul these areas as often as possible during a shift. The increased presence alone led to a reduction of thefts by 12% and a reduction of 26% in the case of burglaries (Cox 2014). Currently, pilot programs are being run in more than 150 cities in the USA. Mohler has created a company specifically for this purpose predpol.com (http://www.predpol.com). Through the police stations concerned, the predictions can be accessed online.

References

Bächle M, Kolb A (2012) Einführung in die Wirtschaftsinformatik. Oldenbourg Wissenschaftsverlag, München

Brantingham J (2014) Predictive policing, UCLA. http://paleo.sscnet.ucla.edu. Accessed 16 Nov 2014

Cox M (2014) BBC Horizon: the age of big data. Gsis Mediacore TV. http://gsis.mediacore.tv/media/bbc-horizon-the-age-of-big-data. Accessed 16 Nov 2014

Desai S, Srivastava A (2013) ERP to E^{2}RP a case study approach. PHI Learning Private Limited, Delhi

Enzyklopädie der Wirtschaftsinformatik (2013) Online-Lexikon. Oldenbourg. http://www.enzyklopaedie-der-wirtschaftsinformatik.de/wi-enzyklopaedie/lexikon/uebergreifendes/Kontext-und-Grundlagen/Informationssystem/Managementunterstutzungssystem. Accessed 16 Nov 2014

Gabriel R, Beier D (2003) Informations management in Organisationen. Kohlhammer, Stuttgart, p S.27

Gartner (2013) Business intelligence (BI), Gartner IT glossary. http://www.gartner.com/it-glossary/business-intelligence-bi. Accessed 16Nov 2014

Gluchowski P et al (2008) Management support systeme und business intelligence. Computergestützte Informationssysteme für Fach- und Führungskräfte, 2. Auflage. Springer Berlin

Grothe M, Gentsch P (2000) Business intelligence – Aus Informationen Wettbewerbsvorteile gewinnen. Addison-Wesley, München u.a.

Hill K (2012) How target figured out a teen girl was pregnant before her father did. Forbes. https://www.forbes.com/sites/kashmirhill/2012/02/16/how-target-figured-out-a-teen-girl-was-pregnant-before-her-father-did/#34ce298c6668. Accessed 03 Apr 2017

Loubriel A (2014) The LAPD is predicting crimes before they happen. Guardian Liberty Voice. http://guardianlv.com/2014/06/the-lapd-is-predicting-crimes-before-they-happen. Accessed 16 Nov 2014

Mohler G (2014) Santa Cruz and Los Angeles predictive policing 6 month Trial. UCLA. http://paleo.sscnet.ucla.edu. Accessed 16 Nov 2014

Plattner H (2013) A course in in-memory data management the inner mechanics of in-memory databases. Springer, Berlin

Swabey P (2013) Tesco saves millions with supply chains analytics. Information Age. http://www.information-age.com/technology/informationmanagement/123456972/tesco-saves-millions-with-supply-chain-analytics. Accessed 16 Nov 2014

Chapter 14
Big Data Analytics

Abstract Big data analytics is introduced as independent but complementary discipline to BI. The focus is on the paradigm shift in terms of entrepreneurial thinking and acting, which was induced by big data. This is supplemented by a detailed discussion about unstructured data, image analytics, text analysis and text mining. The fast parallel processing of large amounts of data independently of their structure leads to Hadoop and MapReduce, with MapReduce being discussed as a technological and methodological basis for Hadoop. Big data analytics process models complete the theoretical part. Twitter text mining, sentiment analysis, visualization and outlier detection conclude this chapter.

14.1 Situation Analysis

Big data is no technology. Big data uses technologies to find the right answers to the right questions in real-time and to implement those methods profitably in daily business. The right combination of big data methods, tools and technologies like Hadoop, in-memory computing, NoSQL databases, social media and traditional data mining enable companies of all sizes to find answers to their questions. The most difficult task for data scientists always is to ask the right questions. Big data is not primarily a data volume problem, but a data complexity problem. This concerns both, the questions as well as access to the right data. Big data requires a customer oriented business model represented by the inverted business pyramid (Plattner 2013) (Fig. 14.1). Linked to this is a transformation of decision making processes based solely on intuition and experience into a data driven process as part of the corporate culture. Big data requires flexibility and a company open to innovation, both from the products and from the business model—and above all, a customer relationship management, which is not a tool, but the basis for an active individual customer relationship. The role of IT changes from the provision and processing of large amounts of data to secure access to data, networks and services, the necessary infrastructure and the provision of big data self services (Gronwald 2014).

Big data analytics is the result of four global trends: Moore's law, mobile computing, social networking, cloud computing (Minelli et al. 2013).

Moore's law postulates (since 1965) that the number of transistors incorporated in a chip will approximately double every 24 months (INTEL 2017) (Fig. 14.2).

© Springer-Verlag GmbH Germany 2017
K.-D. Gronwald, *Integrated Business Information Systems*,
DOI 10.1007/978-3-662-53291-1_14

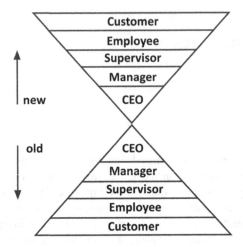

Fig. 14.1 The inverted management pyramid. Source: Adapted from Joan A. Barton and D. Brian Marson, Service quality, Province of Bristich Columbia, 1991

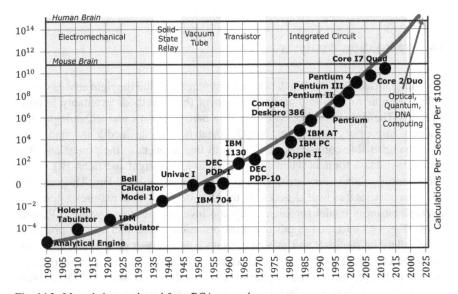

Fig. 14.2 Moore's law—adapted from BCA research

14.2   Big Data Between Business and IT

McAfee and Brynjolfsson (2012) refer to big data as *the management revolution*. They describe big data as a new culture of decision-making. They value management challenges higher than technical issues, especially for the senior executive teams. On the other hand, EMC (2012) puts data scientists on stage as *rockstars of the big data era*. Davenport and Patil (2012) refer to data scientist as *the sexiest job of the twenty-first century.* It would be a mistake, however, to put data scientist on the IT side. Data scientists are analysts, i.e. business people who use technology to solve business problems.

14.3 Paradigm Shift

14.3.1 Separating BI and Big Data

Traditional data mining (BI) segments customers into groups for target group oriented marketing. It is the search for patterns that optimally classify customer groups and develops models to assign customers to these target groups. The customer does not exist in these models as an individual, as a person.

The big data approach uses data mining results for real-time analysis of customer behavior on an individual basis and draws conclusions from group behavior. It is the shift from a product-oriented to the customer-centric cross-selling and up-selling with focus on individual customers as persons.

14.3.2 Case Study Paradigm Shift: The Disney MagicBand

Disney has decades of experience in *crowd management* and *crowd control*, the optimization movements of crowds through its theme parks. Until now, Disney was only able to observe the behavior of a crowd as a single blob that was a single item. MagicBands are bracelets with an RFID sensor, which allows the realtime analysis of the behavior of each individual parking visitor (Fig. 14.3). They combine individual visitor behavior with traditional analytics and group behavior MagicBands allow Disney to study guests as individuals exhibiting unique behavior within the crowd as the crowd moves through the park (Franks 2014).

Example: A child comes to a Disney theme park for the first time. One of the greatest experiences is the direct contact with a princess or Mickey Mouse (Fig. 14.4). While e.g. Mickey approaches the child, the handler for Mickey reads the child's MagicBand with a tablet and brings the information to the screen: This is Jane Smith from Atlanta, she's here for her ninth birthday and she loves gummy bears. The control center analyzes and decides which offer is made to the child, based on the information about her and her family. Now consider how magical it will be for the child when Mickey says "Hi, Jane. It's great to see you here. You came a long way from Atlanta, and I'm so happy that you're celebrating your birthday with us. If you go to that candy store right over there, you can choose a pack of gummy bears as my birthday gift to you. Just tell them I sent you, and they'll hand you your candy with a smile!" If the child's family goes to the store, the cashier will see that the offer of free candy was extended and will quickly process

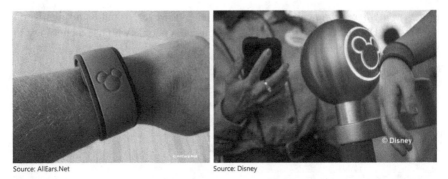

Source: AllEars.Net   Source: Disney

Fig. 14.3 The Disney MagicBand

Source: rentitoday.com   Source: rockerfruit.com

Fig. 14.4 Disney MagicBand—individual customer care

the transaction (Franks 2014). Disney has invested one billion dollars in this system (Garcia 2013).

14.4 The Seven+ Vs

Big data has been characterized by a growing and varying number of *dimensions*, of which only some are applicable to big data exclusively. It started with three dimensions in 2001 already: *Volume, Velocity, Variety* at a time when big data did not even exist nor was it technologically possible and focused mainly on technical aspects of data management in centralized data warehouses (Laney 2001). Most important for big data, however, are the *semantically* related dimensions with *Veracity* appearing as one of the first which considers the credibility of data, e.g. how to recognize sarcasm in sentiment analysis. *Validity, Volatility, and Value* (Normandeau 2013) are completing the most relevant seven. Some authors added *Variability* and *Visualization* (McNulty 2014) which are not considered here. For one, regarding variability at least some aspects have already been covered by

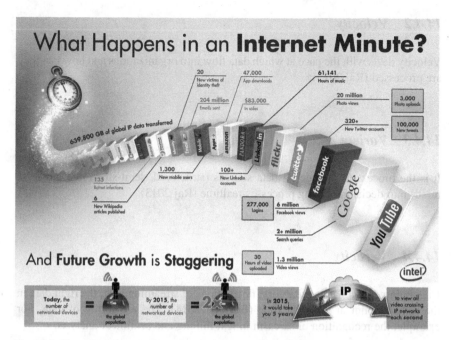

Fig. 14.5 The seven Vs—where the data come from. Source: Hongkiat (2014)

one or more of the other semantic dimensions creating redundancy to some degree. Regarding visualization one would need to add other aspects like *machine learning* which are more related to analytics than serving as dimension to mainly characterize what big data really is and how it differs from traditional business intelligence (Fig. 14.5) (Hongkiat 2014).

14.4.1 Volume

A single jet engine generates 30 tb of data in 30 minutes. With more than 25,000 flights per day, the data volume of a single data source can become some Petabyte (Raj 2013). It is not so much the data volume which characterizes big data, but that these data incur streaming, realtime and must be analyzed immediately, without storing it into a data warehouse first. Only processing of large amounts of data is not new. That existed already before the big data era (Laney 2001).

14.4.2   Velocity

Velocity deals with the pace at which data flow into organizations and how fast they are processed (Raj 2013).

14.4.3   Variety

It is the process to include all data sources (structured, semi-structured, unstructured) into decision making processes realtime (Raj 2013).

14.4.4   Veracity

Uncertainty about data availability, fluctuations of streaming data, the right data in the right amount at the right time (Bowden 2014). The credibility of data. For example, the recognition of sarcasm in text mining and sentiment analysis.

14.4.5   Validity

It is linked to veracity, but has the focus on the correctness and if they are accurate for the intended use, adding a context specific aspect and view to data (Normandeau 2013).

14.4.6   Volatility

Big data volatility refers to how long is data valid and how long should it be stored, if at all. In this world of real time data organizations need to determine at what point is data no longer relevant to the current analysis (Normandeau 2013).

14.4.7   Value

Who benefits directly from the information obtained? Which business decisions have to be made? When is the information needed to make better decisions (Raj 2013)?

14.5   The Problem of Unstructured Data

Big data is the sum of structured and unstructured data from different sources and media that are processed realtime. At this point the data volume becomes important. 90% of all data since the dawn of time were generated between 2012 and 2014. Most of these data come from unstructured sources. These could no longer be processed conventionally, because they could no longer be stored temporarily. With MapReduce, Google has developed a cost-effective method for parallel processing of large amounts of data using clusters of commercial computers. Hadoop is an open source interface to MapReduce. Hadoop has become popular because it can deal with unstructured, semi-structured or quasi-structured data.

Actually, there are no real unstructured data. All data have one or more (hidden) structures. Data are called unstructured when formatted in such a complex way that it cannot be easily transformed into an analytic form (Franks 2014).

In order to identify the optimal procedures and the corresponding tools for processing data, they are divided into classes according to the type of their structure (EMC 2012) (Fig. 14.6):

1. **Structured**

    Data containing a defined data type, format, structure. Example: Transaction data and OLAP.

2. **Semi-Structured**

    Textual data files with a discernable pattern, enabling parsing. Example: XML data files that are self describing and defined by an xml schema.

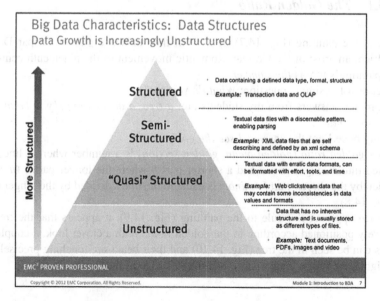

Fig. 14.6 Big data—data structures

3. **Quasi-Structured**

Textual data with erratic data formats, can be formatted with effort, tools, and time. Example: Web clickstream data that may contain some inconsistencies in data values and formats.

4. **Unstructured**

Data that has no inherent structure and is usually stored as different types of files. Example: Text documents, PDFs, images and videos.

14.6   Approach to Image Analytics

Any data, if structured or unstructured have to be converted into a mathematical model in order to be analyzed with big data analytics methods. Data mining as basic big data method is looking for patterns in a huge amount of data. But looking for patterns requires asking the right questions. This applies to any data. Looking for patterns in images is a good example. Face recognition is looking for certain patterns in human faces to identify individuals. Fingerprint identification is doing the same, looking for patterns and so converting unstructured into structured data.

But what's about a painting created by an artist or a portrait photography created by a professional photographer? Can we find patterns there? It begins again with asking the right question.

14.6.1   The Golden Ratio

Look at the painting (Fig. 14.7). It is from the landscape painter Caspar David Friedrich, an artist of the German Romantic movement of the nineteenth century. The painting's title is *the summer*.

Let us ask a simple question: *What do you see?*

Common answers from my students are: *a tree, a dominating sky, a couple, a river, some hills...*

Only once I got the answer: *I see the golden ratio.*

The *golden ratio (golden mean, golden section)* is a number where a line *s* is divided into a longer part *a* and a shorter part *b* where the longer part *a* (*major*) divided by the shorter part *b* (*minor*) is equal to the line s divided by the longer part a (Fig. 14.8).

When applying this rule to the painting (Fig. 14.9), it appears that the tree is precisely positioned according to the golden ratio. With a closer look, a couple of doves can be seen in the tree (Fig. 14.10) and their beaks are touching precisely at the right edge of the longer part *a* and the left edge of the shorter part *b*.

Fig. 14.7 Caspar David Friedrich—the summer—Source: caspardavidfriedrich.org

$$\varphi = \frac{a}{b} = \frac{s}{a} = \frac{a+b}{a} = \frac{1+\sqrt{5}}{2} = 1.618033988 \ldots$$

$$\longleftarrow\!\!\!\!-\!\!\!\!-\!\!\!\!-\!\!\!\!-\!\!\!\!-\!\!\!\!- s -\!\!\!\!-\!\!\!\!-\!\!\!\!-\!\!\!\!-\!\!\!\!-\!\!\!\!\longrightarrow$$

$$a = 0.618033988 \ldots s \qquad\qquad b = 0.381966011 \ldots s$$

Fig. 14.8 The golden ratio

$a = 0.618\,s$ $\qquad\qquad$ $b = 0.382s$

Fig. 14.9 The summer—the golden ratio

Fig. 14.10 The summer—the golden ratio in detail

The golden ratio plays a significant role in human's perception of *beauty*, *harmony*, and *aesthetics* starting with its use in arts, over the ideal proportions of the human body from ancient Greek sculptures up to comic figures (Fig. 14.11). And it can be found in architecture too (Fig. 14.12).

14.6.2   The Golden Ratio and the Fibonacci Numbers

A rectangle with the long side $s = a + b$ (*s = major + minor*) and the short side a (*major*) of the golden ratio is called the *golden rectangle* (Fig. 14.13).

Dividing the long side s at the golden ratio splits the rectangle into a square with the sides major1 and a smaller rectangle which is again a golden rectangle. Splitting the second rectangle at the golden ratio will again result in a square and a golden rectangle. This procedure can be repeated ad infinitum. The (normalized) lengths of the diagonals of the squares are *fibonacci numbers*. Starting with the innermost square and normalize the length of its diagonal to 1 and spiraling the diagonals inside out, the sequence of lengths becomes the *fibonacci sequence*.

In a fibonacci sequence (0, 1, 1, 2, 3, 5, 8, 13, 21, 34, 55 . . .), each term is the sum of the two previous terms. The ratio of a term to the one before will get closer and closer to the golden ratio

$$\frac{a}{b} = \frac{s}{a} = \frac{1 + \sqrt{5}}{2} = 1.618033988\ldots$$

the bigger the fibonacci number is:

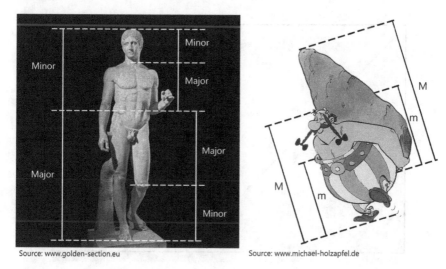

Source: www.golden-section.eu   Source: www.michael-holzapfel.de

Fig. 14.11 The golden ratio and the human body

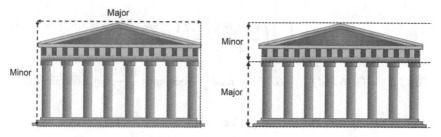

Fig. 14.12 The golden ratio in architecture

$$\text{for } f_7 = 8 \text{ and } f_6 = 5 \text{ the ratio} \frac{f_7}{f_6} \text{ becomes } \frac{8}{5} = 1.6$$

$$\text{for } f_9 = 21 \text{ and } f_8 = 13 \text{ the ration} \frac{f_9}{f_8} \text{ becomes } \frac{21}{13} = 1.61538$$

$$\text{for } f_{11} = 55 \text{ and } f_{10} = 34 \text{ the ratio} \frac{f_{11}}{f_{10}} \text{ becomes } \frac{55}{34} = 1.61764\ldots$$

There is an explicit formula to calculate the nth term of the fibonacci sequence. It is named after the mathematician Jaques Philippe Marie Binet, who derived it 1843 and sometimes includes Abraham de Moivre, who did it already 1718. In between, it was also known to Leonard Euler and Daniel Bernoulli (Weisstein 2013):

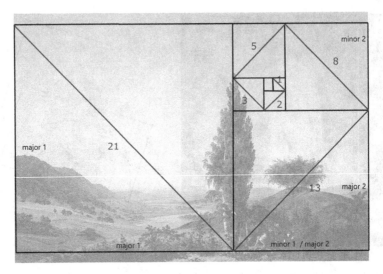

Fig. 14.13 The golden ratio and fibonacci numbers

$$f_n = \frac{1}{\sqrt{5}} \left[\left(\frac{1+\sqrt{5}}{2} \right)^n - \left(\frac{1-\sqrt{5}}{2} \right)^n \right]$$

for $f_{21} = 10946$ and $f_{20} = 6765$ the ratio $\frac{f_{21}}{f_{20}}$ becomes $\frac{10946}{6765} = 1.680339985$

which is already very close to the golden ratio.

14.6.3 The Fibonacci Spiral

Including quarter-circle tangents in the interior of each square from Fig. 14.13 results in a spiral, the *fibonacci spiral* (Fig. 14.14).

With the fibonacci numbers and the golden ratio closely related, it is not surprising that both fibonacci spiral (Fig. 14.15) and the golden ratio (Fig. 14.16) are used as aesthetical design elements in photography.

More important for pattern recognition is the presence of the fibonacci numbers and the fibonacci spiral in nature. That ranges from the shell of the chambered Nautilus up to weather patterns (Fig. 14.17). Obviously, the fibonacci spiral and the fibonacci sequence provide biological advantages to plants (Parveen 2017). Sunflowers have a golden Spiral seed arrangement, because it maximizes the number of seeds that can be packed into a seed head. Branching plants and trees exhibit fibonacci numbers (Adler 1997). This design provides the best physical accommodation for the number of branches, while maximizing sun exposure.

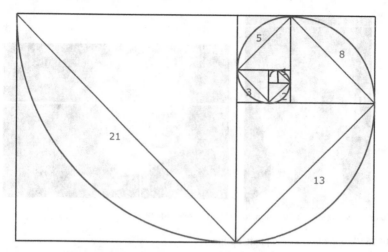

Fig. 14.14 The fibonacci spiral

Fig. 14.15 The fibonacci spiral—source: www.Fotokurs.com

14.7 Visualization: Data Quality—Outlier Detection

14.7.1 Introduction

The assessment of data regarding their quality and validity is of great importance for all statistical and analytical methods. This includes visualizations, that is, the graphical representation of big data results. Looking for patterns is similar to image analytics. In statistics, the normal distribution is a very common continuous probability distribution. Outliers are data that are outside an expected distribution

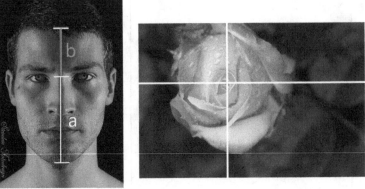

Source: www.steinlein-fotodesign.de   Source: www.digitipps.ch

Fig. 14.16 The golden ratio in photo design

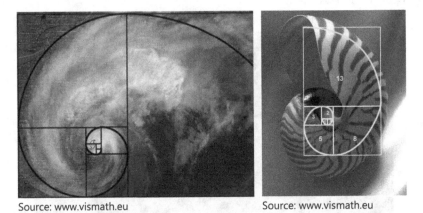

Source: www.vismath.eu   Source: www.vismath.eu

Fig. 14.17 Fibonacci spiral in nature

function. Analysts have to decide on a case-by-case basis, what happens with them, whether they are further considered, ignored or treated separately. However, they must first be identified.

14.7.2   The Normal Distribution (Harrington 2009)

An ordinary dice has a data set $(1, 2, 3, 4, 5, 6)$
The arithmetic mean μ for n $= 6$ is

$$\mu = \frac{1}{n}\sum_{i=1}^{n}x_i = \frac{1+2+3+4+5+6}{6} = 3.5$$

If k dice are thrown once or one dice is thrown k times, the following random values result for $k = 10$:

$$6\ 2\ 5\ 4\ 2\ 3\ 5\ 1\ 1\ 3$$

The arithmetic mean for this random distribution becomes

$$m = \frac{1}{n}\sum_{i=1}^{n}x_i = \frac{6+2+5+4+2+3+5+1+1+3}{10} = 3.2$$

$$\text{for } k \rightarrow \infty \ \ m = \mu$$

The result is shown in Fig. 14.18 as mean value distribution.

20 values of m are between 3.0 and 3.5. The *mode* is the value that appears most often in a set of data. With this resolution, there is more than one mode, that are all between 3.0 and 3.5.

The *median* is the middle score for a set of data that has been arranged in order of magnitude. This is true for an odd number of values. For an equal number of values the median is the average of the middle two scores.

The median for (1,2,3,4,5,6): $3 + 4/2 = 3.5$

The number of experiments will be increased step-by-step. For $N = 500,000$ the results are shown in Fig. 14.19.

The result is the variance of the sample mean values around the arithmetic mean. For the distribution of values with equal probability (dice: 1/6), the values are normal distributed. The resulting curve is the *normal distribution* or *gaussian bell curve* or *gaussian distribution*.

The analytical formula for the gaussian distribution is

$$f(x,\mu,\sigma^2) = \frac{1}{\sigma\sqrt{2\pi}}e^{-\frac{1}{2}\left(\frac{x-\mu}{\sigma}\right)^2}$$

With *variate* x, *mean* μ, *variance* σ^2

The *variate* is defined as the set of all random variables that obey a given probabilistic law.

The square root of the *variance* is the *standard deviation* σ.

$$\sigma = \sqrt{\frac{\sum x^2}{n} - \mu^2}$$

The standard deviation σ is the distance between the mean and the *infliction point* of the normal distribution. In a series of measurements in which the values are

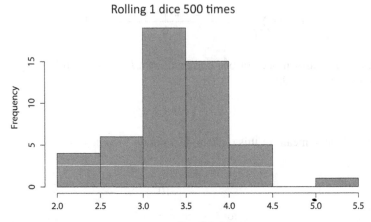

Fig. 14.18 Mean value distribution rolling 1 dice 500 times

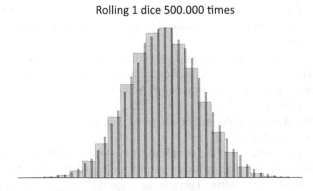

Fig. 14.19 Mean value distribution rolling 1 dice 500,000 times

normal distributed, i.e. have a symmetrical and linear probability distribution, theoretically all mean values are within the bell curve ($-\infty$ to $+\infty$)

At $\pm 1\sigma$ 68% of all mean values are within the interval. At $\pm 6\sigma$ 99.9999998...% of all mean values are within the interval (Fig. 14.20). See Sect. 10.7 *Six Sigma*.

14.7.3 Data Quality and Outlier Detection

Example: Case study from Torgo (2011): *Predicting Algae Bloom*

High concentrations of certain harmful algae in rivers constitute a serious ecological problem with a strong impact not only on river lifeforms, but also on water quality. Being able to monitor and perform an early forecast of algae blooms is essential to improving the quality of rivers. With the goal of addressing this prediction problem, several water samples were collected in different European rivers at different times during a period of approximately 1 year. For each water

Fig. 14.20 Standard
deviation of the normal
distribution

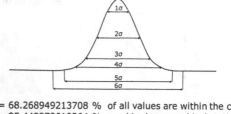

μ ± 1σ = 68.268949213708 % of all values are within the curve
μ ± 2σ = 95.449973610364 % used in demographical statistics
μ ± 3σ = 99.730020393674 %
μ ± 4σ = 99.993665751633 %
μ ± 5σ = 99.999942669685 %
μ ± 6σ = 99.999999802682 % 3.4 defects in 1 million parts

sample, different chemical properties were measured as well as the frequency of occurrence of seven harmful algae. Some other characteristics of the water collection process were also stored, such as the season of the year, the river size, and the river speed. As such, obtaining models that are able to accurately predict the algae frequencies based on chemical properties would facilitate the creation of cheap and automated systems for monitoring harmful algae blooms.

For the following analysis of the data quality, only the pH value is considered. Only the first 200 measured values for the pH-value are considered. The superposition with a normal distribution shows some irregularities (Fig. 14.21).

The data are analyzed using QQ (Quantile-Quantile) plot. The Q-Q plot compares the data series with the normal distribution and identifies outliers outside the confidence interval. The confidence interval is usually 2σ (95%). The central straight line corresponds to the normal distribution; the dashed lines indicate the confidence interval (Fig. 14.22).

Box plots provide a quick summarization of some key properties of the variable distribution. Namely, there is a box whose vertical limits are the 1st and 3rd quartiles of the variable. This box has a horizontal line inside that represents the median value of the variable. Let r be the inter-quartile range. The small horizontal dash above the box is the largest observation that is less than or equal to the 3rd quartile plus $1.5 \times$ r. The small horizontal dash below the box is the smallest observation that is greater than or equal to the 1st quartile minus $1.5 \times$ r. The circles below or above these small dashes represent observations that are extremely low (high) compared to all others, and are usually considered outliers. Outliers are data that are at least 1.5 times outside the interquartile range IQR ($IQR = Q3 - Q1$) (Fig. 14.23).

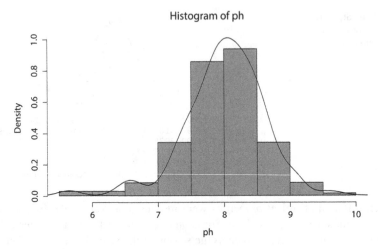

Fig. 14.21 Algae bloom pH measurement distribution

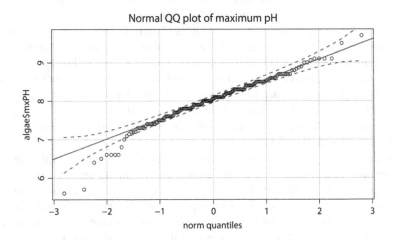

Fig. 14.22 Algae bloom outlier detection with QQ plot

14.8 Text Mining

Text mining is primarily the process of preparing unstructured text in such a way that it can be further treated with other analytical methods in order to obtain information from it. Text analytics includes

- Content categorization: classification of text documents into categories,
- Text mining: recognizing patterns and structures and making predictions or understanding the behavior,
- Sentiment analysis: assessment of text content as positive or negative (polarization).

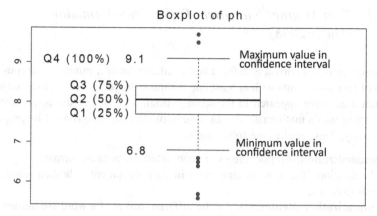

Fig. 14.23 Algae bloom outlier detection with boxplot

14.8.1 Text Mining: Categories

Text mining can be subdivided into seven categories (Miner 2012):

1. Search and information retrieval: Storing and retrieving text documents (e.g., search engines, keyword search).
2. Document clustering: Grouping and categorizing terms, sections, paragraphs, and documents using data mining clustering methods.
3. Document classification: Grouping and classification of sections, text passages, paragraphs and documents using data mining classification methods.
4. Web mining: Data and text mining in the web.
5. Information extraction: Identification and extraction of relevant facts and relationships from unstructured text, which involves transforming unstructured and semi-structured texts into structured data.
6. Natural language processing: Man-machine interaction, which allows computers to determine meaning and natural language and to derive actions from it (computational linguistics).
7. Concept Extraction: Grouping of words and sentences into semantically similar groups.

14.8.2 Text Mining: Linguistic and Mathematical Approach

There are two basic approaches to text mining, the linguistic approach, the attempt to determine structure and meaning through grammatical rules, and the mathematical approach of numerical methods to extract as much information from texts as possible. The mathematical approach requires several steps to transform text data into a numerical form, which is understood by mathematical, analytical methods. Sentiment analysis is understood as a linguistic approach (Duffy 2008).

14.8.3 Text Mining: Numerical Duffy Transformation (Duffy 2008)

It is assumed that a corpus is defined and available for text mining. A corpus is a group of text documents with at least one common property (belong to a common research area, have appeared in the same journal, ...). Before the corpus can be transformed into a mathematical data mining model, the corpus must be prepared accordingly. This requires the following steps:

1. **Standardization**: All documents are converted to the same format.
2. **Tokenization**: The flow of characters in each document is broken down into words (tokens).
3. **Lemmatization (stemming)**: All the different forms of a word are unified and brought into their basic form (for example, transformation of all plural forms into their singular form).
4. **Dictionary reduction (stop word removal)**: Stop words are words that very rarely include predictability such as articles and pronouns.
5. **Vector generation**: Documents are represented as vectors, each word root representing one line. Each document is a column. This results in a matrix which is also used as a vector space model of the corpus for semantic analysis. In many cases the frequency of a word is also used instead of the word itself (frequency scoring). This can be normalized since the relative frequency of a word in the entire body is more important than the absolute frequency in a text.

14.8.4 Text Mining: Numerical Lu Transformation (Lu 2013)

Lu avoids the standardization step and instead inserts a filter step that removes special characters and punctuation. This is of greater importance for websites. In addition, he introduces *pruning*, which removes words with a very low frequency. Lu calls the generated vector-space model *bag of words* or *dictionary model*. This is used by search engines that compare the similarity between the terms of a document and the words of a query. There are several other authors who modify this model and adapt it to their specific needs.

14.8.5 Text Mining: Vector Space Model

The vector space model goes back to Gerard Salton, who developed it in the 1960s under the SMART system. SMART stands for *System for the Mechanical Analysis and Retrieval of Text* (Salton 1968).

In a vector space model, a vector represents a term (concept, keyword, term) that belongs to a particular document. A weighting value associated with a term represents the importance of the term with respect to the semantics of the document.

Example: Analysis of documents dealing with *beer market development*. In the database (Corpus) these are stored as indicated terms *beer, market, development*.

A query q with the three terms can be represented as a vector in the three corresponding dimensions. The same applies to documents d, which are to be examined for the occurrence of these terms. The terms are weighted according to certain criteria in both the query q and the documents d (which will be discussed later). The degree of correspondence between the query q and the document d results from the distance between the documents in euclidean space, that is, from the angle between both vectors $\vec{q}$ and $\vec{d}$. This is calculated as an inner product between $\vec{q}$ and $\vec{d}$ (Fig. 14.24).

$$\cos \alpha = \frac{\vec{q} \cdot \vec{d}}{\|\vec{q}\| \cdot \|\vec{d}\|}$$

$\|\vec{q}\|$ is the norm of vector $\vec{q}$, $\|\vec{d}\|$ is the norm of vector $\vec{d}$

The norm of the vector $\vec{q}$ is calculated as

$$\|\vec{q}\| = \sqrt{\sum_{i=1}^{n} q_i^2}$$

The norm of the vector $\vec{d}$ is calculated as

$$\|\vec{d}\| = \sqrt{\sum_{i=1}^{n} d_i^2}$$

with $\vec{q} \rightarrow = \begin{pmatrix} 0.8 \\ 1.0 \\ 0.6 \end{pmatrix}$ and $\vec{d} \rightarrow = \begin{pmatrix} 1.0 \\ 0.8 \\ 0.8 \end{pmatrix}$ cos α becomes

$$\cos \alpha = \frac{0.8*1.0 + 1.0*0.8 + 0.6*0.8}{\sqrt{0.8^2 + 1.0^2 + 0.6^2} * \sqrt{1.0^2 + 0.8^2 + 0.8^2}}$$

$$= \frac{2.08}{\sqrt{2} * \sqrt{2.28}} = \frac{2.08}{2.14} = 0.97$$

At an angle of zero degrees the cosine equals one. This gives the greatest alignment between the query and the document. At an angle of ninety degrees the cosine equals to zero and there is no correspondence between the document and the query. In the same way, documents can be compared to each other for similarities.

For i documents D_i, a specific query Q_j and t terms T_k cos(D_i, Q_j)becomes

Fig. 14.24 Text mining—vector space model

$\vec{q} = (0.8, 1.0, 0.6)$

$\vec{d} = (1.0, 0.8, 0.8)$

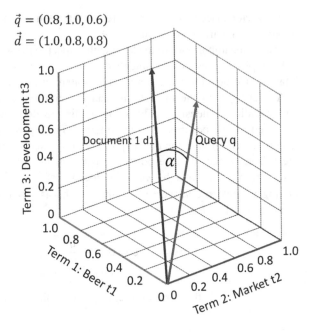

$$\cos\left(D_i, Q_j\right) = \frac{\sum\limits_{k=1}^{t}\left(T_{ik}\cdot Q_{jk}\right)}{\sqrt{\sum\limits_{k=1}^{t}\left(T_{ik}\right)^2\cdot\sum\limits_{k=1}^{t}\left(Q_{jk}\right)^2}}$$

There are a couple of different methods and variants for the weighting of terms. The most common is the *tf-idf* weighting.

The *tf* factor (term frequency factor) describes how frequently the term T_k occurs in a document D_j. It is assumed that the frequency of the occurrence of a term in a document is a measure of the importance of the document.

The *idf* factor (inverse document frequency) is a measure for the number of documents with the term T_k.

Thus, the weight $w_{i,k} = tf_{i,k}\cdot idf_k$.

As a result of the numerical transformation, the corpus contains the number d of documents described by the terms t. This is shown as

$$T \times D \quad Matrix\ A$$

The document vectors D_i are the rows, the term vectors D_k are the columns of the matrix.

For six documents and three terms the matrix becomes

$$A = \begin{pmatrix} T_{11} & T_{12} & T_{13} \\ T_{21} & T_{22} & T_{23} \\ T_{31} & T_{32} & T_{33} \\ T_{41} & T_{42} & T_{43} \\ T_{51} & T_{52} & T_{53} \\ T_{61} & T_{62} & T_{63} \end{pmatrix}$$

Example for Beer Market Development
Terms:

T_1: beer
T_2: market
T_3: development

Documents:

D_1: $\cos \alpha = 0.9848$—The *development* of the *beer market* in the coming year
D_2: $\cos \alpha = 0.7071$—The *market* is stagnating due to higher *beer* prices
D_3: $\cos \alpha = 0.4694$—The *beer market* is declining
D_4: $\cos \alpha = 0.9659$—*Market* analyzes predict a good *development* for *beer*
  consumption
D_5: $\cos \alpha = 0.3420$—Everything you want to know about *beer*
D_6: $\cos \alpha = 0.5735$—*Beer* cans have little chance in the *market*

This results in the following matrix without weighting:

$$A = \begin{pmatrix} 1 & 1 & 1 \\ 0 & 1 & 1 \\ 1 & 0 & 1 \\ 1 & 1 & 1 \\ 0 & 0 & 1 \\ 1 & 0 & 1 \end{pmatrix}$$

With weighting:

$$A = \begin{pmatrix} 0.9848 & 0.9848 & 0.9848 \\ 0 & 0.7071 & 0.7071 \\ 0.4694 & 0 & 0.4694 \\ 0.9659 & 0.9659 & 0.9659 \\ 0 & 0 & 0.3420 \\ 0.5735 & 0 & 0.5735 \end{pmatrix}$$

14.9   MapReduce: Basics

With *MapReduce* Google has developed a cost-effective method for parallel processing of large amounts of data using clusters of commercial computers. *Hadoop* is an open source interface to MapReduce. MapReduce stands for the processing of big data and extracts only the data that are needed. The data sets of Big Data are too large to be processed using conventional methods. They simply cannot be further scaled. MapReduce is based on the principle of parallelism (Nahrstedt and King 2007). It achieves the speed of processing large amounts of data by parallel processing of data combined with data reduction. These processes can be performed recursively.

A prerequisite for parallelism is that data and tasks are interchangeable without changing their result (commutative law).

Example: $x = A + B$ with $A = (a*b)$ and $B = (y*z)$

$$A + B = B + A$$

Thus, A and B can be processed independently in parallel. This is called *data parallelism*. (Nahrstedt and King 2007).

Dividing work into larger tasks identifies logical units for parallelization as threads is called *task parallelism* These can commutate internally, whereby in some places of the process they must be synchronized, that is, do not commutate and internal parallelism is not fully exploited (Fig. 14.25). Intelligent task design eliminates as many synchronization points as possible (Nahrstedt and King 2007).

Hicks (2013) distinguishes between three kinds of parallelism:

- Data parallelism—the same task run on different data in parallel.
- Task parallelism—different tasks running on the same data.
- Hybrid data/task parallelism—a parallel pipeline of tasks, each of which might be data parallel.

Map is the first step in each MapReduce process. In the mapping phase, the raw data is transformed into vectors (key, value), e.g. the key can be a row number and the value is a text string. The map function transforms this input into a series of output pairs. In this case, a record containing each occurring word (key) and the frequency of its occurrence (value). Example: In a document D_i, there is the sentence

$$Sentence_{k=1} = to\ be\ or\ not\ to\ be$$

Map splits the sentence into words as keys and the number of occurrences as values

Task A   Task B

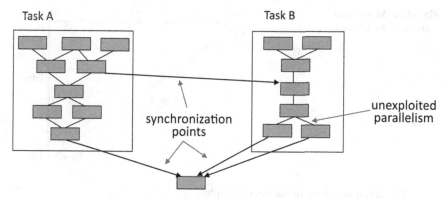

synchronization
points

unexploited
parallelism

Fig. 14.25 MapReduce task parallelism (Nahrstedt and King 2007)

$$key_{i1} = [[to, 2], [be, 2], [or, 1], [not, 1]]$$

This process will be repeated and processes in parallel for additional documents D_i.

Reduce aggregates all occurrences k of key_{i1} to idealy one final key $key1$

$$key1 = [[be, 48], [or, 24], [not, 24]]$$

14.10   MapReduce Application: Social Triangle, e-Discovery (EMC 2012)

Example: Suppose there are 517,424 emails from an energy company under indictment. The potential fraud of the manager Walt is investigated. To find out who additional might be involved, the social network of Walt is investigated. A social network is a group of people who regularly communicate with each other. More than 500,000 emails are being investigated. This is done in three steps with the help of a social graph, which represents the personal relationships of the members of this network regarding the e-mail traffic (Fig. 14.26).

14.10.1   Step 1: Edge 1

Mapper 1

- maps two regular expression searches:
    From: Walt
    To: Michael, Dan, Lori, Susan

Fig. 14.26 MapReduce
e-discovery social graph

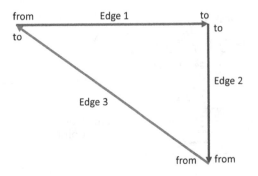

The result is Edge1 in the social graph:
[*key, value*] = [*Walt, (Michael, Dan, Lori, Susan)*]

Reducer 1

- gets the output from the mapper with different values:
    [*key, value*] = [*Walt, (Michael, Dan, Lori, Susan)*]
    [*key, value*] = [*Walt, (Lori, Susan, Jeff, Ken)*]
- unions the values for the second directed edge:
    [*key, value*] = [*Walt (Dan, Jeff, Ken, Lori, Michael, Susan)*]

Result: Data is reduced by about 30%.

14.10.2 Step 2: Edge 2

Mapper 2

- reverses the previous Map:
    To: Michael, Dan, Lori, Susan
    From: Walt
    Emits the inbound directed edge of the social graph:
    [*key, value*] = [*Susan, Walt*]; [*Lori, Walt*]; [*Dan, Walt*];...

Reducer 2

- gets the output from the mapper with different values
    [*key, value*] = [*Susan, Walt*]
    [*key, value*] = [*Susan, Jeff*]
- unions the values for Edge 3:
    [*key, value*] = [*Susan (Jeff, Ken, Walt)*]

Data again reduced by about 30%.

14.10.3 Step 3: Edge 3

Mapper 3

- joins [*inbound*] and [*outbound*] lists by key
    [*key, value*] = [*Walt*, ((*Jeff, Ken, Lori, Susan*), (*Jeff, Lori, Stanley, Walt*))]
- emits [person, person] pair with level of association:
    [*key, value*] = [*Walt: Susan, reciprocal*]; [*Walt: Lori, directed*];...

Reducer 3

- unions the output of the mappers and presents rules:
    [*key, value*] = [*Walt: Susan, reciprocal*]
    [*key, value*] = [*Walt: Lori, directed*]

The Mapper 3 takes the combination of inbound (mail sent to) and outbound (mail sent from) and outputs a key value pair in the form <Sender: Recipient, relationship>. In this instance the relationships are defined as *reciprocal* (Sender sent/received mail to/from the recipient) or *directed* (person sent mail to the recipient). The third reducer can shape the data any way that serves the business objective.

14.11 Hadoop

The term *Hadoop* is often used as a synonym for representing THE big data tool. First, Hadoop is nothing more than a front end to MapReduce, which is well suited for the combined processing of structured and unstructured data.

Hadoop is used as synonym for a couple of additional meanings (EMC 2012):

- Description of the MapReduce Paradigm.
- Massive unstructured data storage on commodity hardware.
- Java Classes for HDFS types and MapReduce job management.
- HDFS: The Hadoop distributed file system.

Hadoop main components (EMC 2012):

- HDFS—Hadoop Distributed File System

  - reliable, redundant, distributed file system optimized for large files.

- MapReduce für Big Data Analytics

  - Programming model for processing sets of data.
  - Mapping inputs to outputs and reducing the output of multiple mappers to one (or a few) answer(s).

Hadoop operational modes (EMC 2012):

- Java MapReduce Mode

  - Write Mapper, Combiner, Reducer functions in Java using Hadoop Java APIs.
  - Read records one at a time.

- Streaming Mode

  - Uses *nix pipes and standard input and output streams.
  - Can be any language (Python, Ruby, C, Perl, Tcl/Tk, etc.).
  - Input can be a line at a time, or a stream at a time.

Hadoop has a set of query languages that are arranged around the core (MapReduce and HDFS) and support the development and manipulation of Hadoop clusters (EMC 2012):

- **Pig**

  - is a data flow language and execution environment for Hadoop,
  - uses the script language Pig Latin for creating Hadoop MapReduce programs.

- **Hive**

  - is a query language based on SQL for building MapReduce jobs,
  - provides web, server and shell interfaces for clients,
  - all data stored in tables.

- **HBase**

  - is a *column oriented* database built over HDFS supporting MapReduce and point queries,
  - depends on *Zookeeper (provides a centralized infrastructure and services that enable synchronization across a cluster)* for consistency and Hadoop for distributed data,
  - the Siteserver component provides several interfaces to Web clients.

Starting with Pig over Hive to HBase, there is an increasing abstraction from an original Hadoop view to an RDBMS view (EMC 2012).

Criteria for selecting the best interface to Hadoop (Pig, Hive, HBase) (EMC 2012):

- **Pig**—Replacement for MapReduce Java coding.
- **Hive**—Apply when there are SQL skills.
- **HBase**—Useful if many different queries are used.

14.12 Analytics Lifecycle: Big Data Analytics Process Models

14.12.1 Initial Situation

Big Data Analytics uses mostly traditional methods of analytics. This also applies to big data analytics process models. Franks (2014) proposes a new model (Big

Data Discovery) but at the same time emphasizes that it is mainly a modified description and therefore modified content of each phase, but the phase model remains largely the same with the same goals to be achieved. Not to be confused with the Oracle Big Data Discovery Tool (Henschen 2014), which serves as Business Self Service Tool and Visualization Interface to Hadoop, similar to e.g. Tableau (http://www.tableausoftware.com). The term Big Data Process Model is therefore used as differentiator. There are two main methods: CRISP-DM (Cross Industry Standard Process for Data Mining) and SAS SEMMA (Sample, Explore, Modify, Model, Assess).

SEMMA and CRISP-DM differ mainly in one phase. SEMMA does not have a business understanding phase as part of the process model, but rather assumes that the business problem is known and solved and described at the beginning of the project (Table 14.1) (Fig. 14.27).

Table 14.1 Big data analytics process models

CRISP-DM	Big data discovery	SAS SEMMA
Business understanding	Analytics idea	Business problem (assumed)
Data understanding	Data loading and integration	Sample and explore
Data preparation		Modify
Modeling	SQL and non-SQL analysis	Model
Evaluation	Evaluation of results	Assess
Deployment	Operationalizing	Deployment (follows)

Source: Franks 2014, The analytics revolution, Wiley

Fig. 14.27 Generic analytics process flow (Franks 2014; EMC 2012)

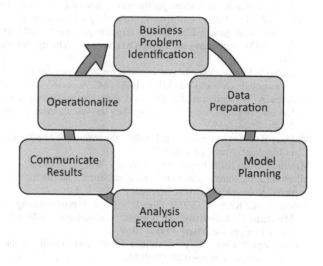

References

Adler I et al (1997) A history of the study of phyllotaxis. Ann Bot 80(3):231–244

Bowden J (2014) The 4 V's in big data for digital marketing. Business 2 Community. http://www.business2community.com/digital-marketing/4-vs-big-data-digital-marketing-0914845. Accessed 16 Nov 2014

Davenport T, Patil D (2012) Data scientist: the sexiest job of the 21st century. Harvard Business Review. https://hbr.org/2012/10/data-scientist-the-sexiest-job-of-the-21st-century. Accessed 16 Nov 2014

Duffy V (2008) Handbook of digital human modeling: research for applied ergonomics and human factors engineering. Taylor & Francis. http://books.google.ch/books?id=Ira9qiakiTMC. Accessed 16 Nov 2014

EMC (2012) Data science and big data analytics course. EMC Corporation. https://education.emc.com/guest/campaign/data_science.aspx. Accessed 16 Nov 2014

Franks B (2014) The analytics revolution. Wiley, New York

Garcia J (2013) Disney's $1 billion wristband project is most expensive in theme park history. skift.com. http://skift.com/2013/08/18/disneys-1-billion-wristband-project-is-most-expensive-in-theme-park-history. Accessed 16 Nov 2014

Gronwald K (2014). Big data und KMU passen zusammen. Swiss IT Magazine 2014 9:6–7

Harrington J (2009), Wahrscheinlichkeit und Normalverteilung. Universität München. http://www.phonetik.uni-muenchen.de/~jmh/lehre/sem/ss09/stat/normal.pdf. Accessed 16 Nov 2014

Henschen D (2014) Oracle unveils Hadoop data exploration tool. InformationWeek, UMB Tech. http://www.informationweek.com/big-data/big-data-analytics/oracle-unveils-hadoop-data-exploration-tool/d/d-id/1316198. Accessed 30 Dec 2014

Hicks M (2013) Distributed programming with MapReduce. http://www.cs.umd.edu/class/fall2013/cmsc433/lectures/mapreduce.pdf. Accessed 4 Apr 2017

Hongkiat (2014) What happens in an internet minute? Hongkiat.com. http://www.hongkiat.com/blog/what-happens-in-an-internet-minute-infographic. Accessed 16 Nov 2014

Intel (2017) Moore's law and intel innovation. Intel Corporation. http://www.intel.com/content/www/us/en/history/museum-gordon-moore-law.html. Accessed 16 May 2017

Laney D (2001) 3D data management: controlling data volume, velocity, and variety, application delivery strategies. META Group. http://blogs.gartner.com/doug-laney/files/2012/01/ad949-3D-Data-Management-Controlling-Data-Volume-Velocity-and-Variety.pdf. Accessed 4 Apr 2017

Lu Z (2013) Information retrieval methods for multidisciplinary applications. IGI Global. http://books.google.ch/books?id=8HqYgBq81_AC. Accessed 16 Nov 2014

McAffee A, Brynjolfsson E (2012) Big data: the management revolution. Harvard Business Review. https://hbr.org/2012/10/big-data-the-management-revolution/ar. Accessed 16 Nov 2014

McNulty E (2014) Understanding big data: the seven V's. http://dataconomy.com/2014/05/seven-vs-big-data/. Accessed 4 Apr 2017

Minelli M et al (2013) Big data big analytics. Wiley, Hoboken

Miner G (2012) Practical text mining and statistical analysis for non-structured text data applications. Academic, Waltham, MA [u.a.]

Nahrstedt K, King S (2007) Google's parallel programming model and implementation. MapReduce. University of Illinois. www.cs.uiuc.edu/class/fa07/cs423/Lectures/Paral02-mapreduce.ppt. Accessed 30 Dec 2014

Normandeau K (2013) Beyond volume variety and velocity is the issue of big data veracity. http://insidebigdata.com/2013/09/12/beyond-volume-variety-velocity-issue-big-data-veracity/. Accessed 4 Apr 2017

Parveen N (2017) Fibonacci in nature. The University of Georgia. http://jwilson.coe.uga.edu/emat6680/parveen/fib_nature.htm. Accessed 4 Apr 2017

Plattner H (2013) A course in in-memory data management the inner mechanics of in-memory databases. Springer, Berlin

Raj S (2013) Big data – an introduction. Amazon. http://www.amazon.com/dp/B00AZCK0Y4?tag=kiq-free-r-20. Accessed 16 Nov 2014

Salton G (1968) Automatic information organisation and retrieval. McGraw-Hill, New York

Torgo L (2011) Data mining with R learning with case studies. CRC Press/Taylor & Francis Group

Weisstein E (2013) Binet's Fibonacci number formula. MathWorld – a Wolfram web resource. http://mathworld.wolfram.com/BinetsFibonacciNumberFormula.html. Accessed 16 Nov 2014

Part III
Information Material

Abstract

This part contains detailed information material that is needed for the decision-making process for the different game rounds. The material is complete and independent of the online game. Templates and other information material can be downloaded from the kdibis server at any time and is not included here. That includes user manuals and the supervisor manual (Chap. 2).

Chapter 15
Post-Merger Situation: Alpha Beer

Abstract The chapter contains the necessary information about the market situation, distribution, customer structure and product portfolio as well as the IT infrastructure for each division directly after the merger.

15.1 Alpha Beer Market: Distribution Structure— Customers

Figure 15.1 shows the Alpha Beer product portfolios for each retail chain after the merger.

Previous year results (Figs. 15.2, 15.3, and 15.4).

15.1.1 Alpha Beer Retailer 1

The retail chain 1 supplies its own stores and large retail chains with the entire portfolio mostly except kegs. In addition to beverages in their own shops, other goods, especially snacks are offered.

Orders from individual stores are collected and sent to the distributor once per week depending on demand and inventory.

The goal is to minimize inventory, while at the same time ensuring delivery. There is a central warehouse supporting all shops. The distributor delivers exclusively to the central warehouse.

There is no demand planning, but a high expectation to the distributor for immediate and complete delivery.

15.1.2 Alpha Beer Retailer 2

Retail chain 2 supplies its own beverage markets with the entire portfolio except single bottles. Cases and kegs are delivered to private customers and restaurants only.

© Springer-Verlag GmbH Germany 2017 161
K.-D. Gronwald, *Integrated Business Information Systems*,
DOI 10.1007/978-3-662-53291-1_15

Fig. 15.1 Alpha Beer product portfolio

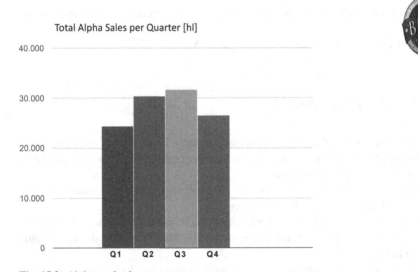

Fig. 15.2 Alpha total sales per quarter

Orders from individual beverage markets are collected and sent to the distributor once per week depending on demand and inventory.

The goal is to minimize inventory, while at the same time ensuring delivery. There is a central warehouse supporting all shops. The distributor delivers exclusively to the central warehouse.

There is no demand planning, but a high expectation to the distributor for immediate and complete delivery.

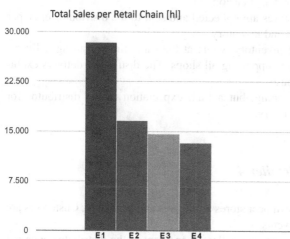

Fig. 15.3 Alpha total sales per retail chain

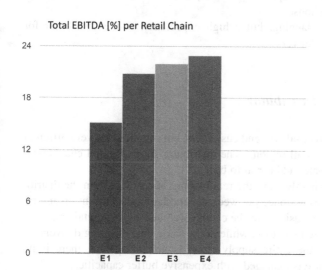

Fig. 15.4 Alpha EBITDA per retail chain

15.1.3   Alpha Beer Retailer 3

The retail chain 3 operates its own beer boutiques in malls and in prominent shopping streets with its own brands. It delivers to private customers only. Cases and kegs are delivered to customers for special private or company events, but are only available on request for select customers.

Orders from individual stores are collected and sent to the distributor once per week depending on demand and inventory.

The goal is to minimize inventory, while at the same time ensuring delivery. There is a central warehouse supporting all shops. The distributor delivers exclusively to the central warehouse.

There is no demand planning, but a high expectation to the distributor for immediate and complete delivery.

15.1.4   Alpha Beer Retailer 4

Retail chain 4 supplies its own beer stores with cases and kegs only. Customers are restaurants, clubs and big events.

Orders from individual stores are collected and sent to the distributor once per week depending on demand and inventory.

The goal is to minimize inventory, while at the same time ensuring delivery. There is a central warehouse supporting all shops. The distributor delivers exclusively to the central warehouse.

There is no demand planning, but a high expectation to the distributor for immediate and complete delivery.

15.1.5   Alpha Beer Distributor

The distributor has no direct sales to end customers, but supplies four very different retail chains exclusively in all regions. The distributor supplies retail chains with soft drinks and bottled water additional to beer.

Orders are accepted directly from the retail chains and orders from the distributor are sent to the wholesaler once per week depending on the current inventory. The distributor delivers exclusively to the central warehouse of the retailers.

The goal is to minimize inventory, while at the same time ensuring delivery.

Since the demand of the entire supply chain is not transparent, there is no forecast, but the warehouse is managed with expensive buffer capacities.

15.1.6   Alpha Beer Wholesaler

The wholesaler has no direct sales to end customers, but supplies the distributor exclusively. The wholesaler supplies distributors with soft drinks and bottled water additional to beer.

Orders are accepted directly from the distributor and orders from the wholesaler are sent to the factory once per week depending on the current inventory. The wholesaler delivers exclusively to the central warehouse of the distributor.

The goal is to minimize inventory, while at the same time ensuring delivery.

Since the demand of the entire supply chain is not transparent, there is no forecast, but the warehouse is managed with expensive buffer capacities.

15.1.7   Alpha Beer Factory

The factory has no direct sales to end customers, but supplies the wholesaler exclusively. It provides the entire portfolio of beer products only.

Orders are accepted directly from the wholesaler and orders from the factory are sent to the production once per week depending on the current inventory.

The goal is to minimize inventory, while at the same time ensuring delivery.

Since the demand of the entire supply chain is not transparent, there is no forecast, but the warehouse is managed with expensive buffer capacities. The factory delivers exclusively to the central warehouse of the wholesaler. The production has unlimited capacity and can deliver demand on a weekly basis. Unexpected additional demand of the factory can be fulfilled within 7 days.

15.2   Alpha Beer IT: Infrastructure

15.2.1   Alpha Beer Retailer 1

There is no order management, no warehouse administration and no human resource management, only a PC-based financial system for processing invoices.

Orders are retrieved from the stores by telephone, whereby each sales person manages these data in a separate PC.

The orders are bundled and then sent directly to the distributor by email.

There is no transparency and there are no systems for demand planning and forecasting and no consistent centralized processes for the retail shops.

15.2.2   Alpha Beer Retailer 2

There is no order management, no warehouse administration and no human resource management, only a PC-based financial system for processing invoices.

Orders are retrieved from the markets by telephone, whereby each sales person manages these data in a separate PC.

The orders are bundled and then sent directly to the distributor by email.

There is no transparency and there are no systems for demand planning and forecasting and no consistent centralized processes for the markets.

15.2.3   Alpha Beer Retailer 3

There is no order management, no warehouse administration and no human resource management, only a PC-based financial system for processing invoices.

Orders are retrieved from the stores by telephone, whereby each sales person manages these data in a separate PC.

The orders are bundled and then sent directly to the distributor by email.

There is no transparency and there are no systems for demand planning and forecasting and no consistent centralized processes for the shops.

15.2.4   Alpha Beer Retailer 4

There is no order management, no warehouse administration and no human resource management, only a PC-based financial system for processing invoices.

Orders are retrieved from the stores by telephone, whereby each sales person manages these data in a separate PC.

The orders are bundled and then sent directly to the distributor by email.

There is no transparency and there are no systems for demand planning and forecasting and no consistent centralized processes for the stores.

15.2.5   Alpha Beer Distributor

There is a separate order management system, a small warehouse management system and a human resource management system, integrated into a PC-based financial system.

Orders are retrieved from the retailers by email. They are synchronized manually with the inventory and order to the wholesaler is entered using the factory's online portal.

There is no transparency and there are no systems for demand planning and forecasting and no consistent centralized processes order management and inventory management.

15.2.6   *Alpha Beer Wholesaler*

There is a separate order management system, a small warehouse management system and a human resource management system, integrated into a PC-based financial system.

Orders are retrieved from the distributor through the factory's online portal. They are synchronized manually with the inventory and orders to the factory is entered using the factory's online portal.

There is no transparency and there are no systems for demand planning and forecasting and no consistent centralized processes order management and inventory management.

15.2.7   *Alpha Beer Factory*

Production is largely automated. There is an ERP system for production planning, finance, human resource management (HR) and customer relationship management (CRM). CRM is used exclusively for order management.

Orders from the wholesaler are placed directly through the factory's online portal.

Supply chain management (SCM) was not considered necessary so far.

The ERP system is 8 years old and a major upgrade is planned, but is on hold due to the merger.

Chapter 16
Post-Merger Situation: Green Beer

Abstract The chapter contains the necessary information about the market situation, distribution, customer structure and product portfolio as well as the IT infrastructure for each division directly after the merger.

16.1 Green Beer Market: Distribution Structure— Customers

Figure 16.1 shows the Green Beer product portfolios for each retail chain after the merger.

Previous year results (Figs. 16.2, 16.3, and 16.4).

16.1.1 Green Beer Retailer 1

The retail chain 1 supplies its own stores and large retail chains with the entire portfolio mostly except kegs. In addition to beverages in their own shops, other goods, especially snacks are offered.

Orders from individual stores are collected and sent to the distributor once per week depending on demand and inventory.

The goal is to minimize inventory, while at the same time ensuring delivery. There is a central warehouse supporting all shops. The distributor delivers exclusively to the central warehouse.

There is no demand planning, but a high expectation to the distributor for immediate and complete delivery.

16.1.2 Green Beer Retailer 2

Retail chain 2 supplies its own beverage markets with the entire portfolio except single bottles. Cases and kegs are delivered to private customers and restaurants only.

© Springer-Verlag GmbH Germany 2017
K.-D. Gronwald, *Integrated Business Information Systems*,
DOI 10.1007/978-3-662-53291-1_16

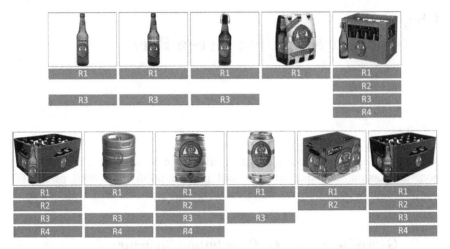

Fig. 16.1 Green Beer product portfolio

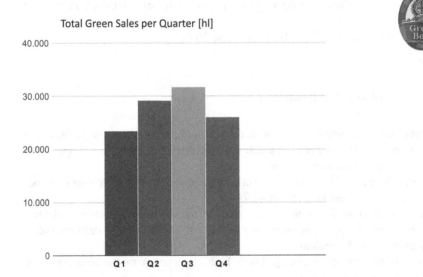

Fig. 16.2 Green total sales per quarter

Orders from individual beverage markets are collected and sent to the distributor once per week depending on demand and inventory.

The goal is to minimize inventory, while at the same time ensuring delivery. There is a central warehouse supporting all shops. The distributor delivers exclusively to the central warehouse.

There is no demand planning, but a high expectation to the distributor for immediate and complete delivery.

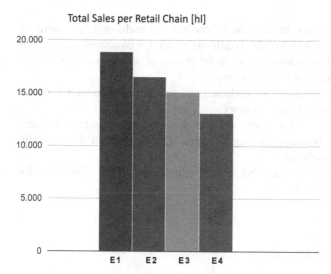

Fig. 16.3 Green total sales per retail chain

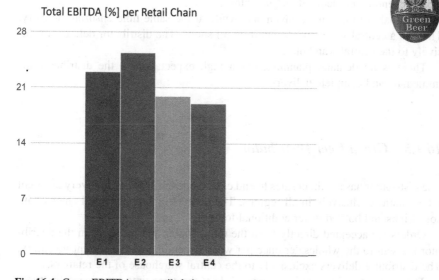

Fig. 16.4 Green EBITDA per retail chain

16.1.3   Green Beer Retailer 3

The retail chain 3 operates its own beer boutiques in malls and in prominent shopping streets with its own brands. It delivers to private customers only. Cases and kegs are delivered to customers for special private or company events, but are only available on request for select customers.

Orders from individual stores are collected and sent to the distributor once per week depending on demand and inventory.

The goal is to minimize inventory, while at the same time ensuring delivery. There is a central warehouse supporting all shops. The distributor delivers exclusively to the central warehouse.

There is no demand planning, but a high expectation to the distributor for immediate and complete delivery.

16.1.4   Green Beer Retailer 4

Retail chain 4 supplies its own beer stores with cases and kegs only. Customers are restaurants, clubs and big events.

Orders from individual stores are collected and sent to the distributor once per week depending on demand and inventory.

The goal is to minimize inventory, while at the same time ensuring delivery. There is a central warehouse supporting all shops. The distributor delivers exclusively to the central warehouse.

There is no demand planning, but a high expectation to the distributor for immediate and complete delivery.

16.1.5   Green Beer Distributor

The distributor has no direct sales to end customers, but supplies four very different retail chains exclusively in all regions. The distributor supplies retail chains with soft drinks and bottled water additional to beer.

Orders are accepted directly from the retail chains and orders from the distributor are sent to the wholesaler once per week depending on the current inventory. The distributor delivers exclusively to the central warehouse of the retailers.

The goal is to minimize inventory, while at the same time ensuring delivery.

Since the demand of the entire supply chain is not transparent, there is no forecast, but the warehouse is managed with expensive buffer capacities.

16.1.6   Green Beer Wholesaler

The wholesaler has no direct sales to end customers, but supplies the distributor exclusively. The wholesaler supplies distributors with soft drinks and bottled water additional to beer.

Orders are accepted directly from the distributor and orders from the wholesaler are sent to the factory once per week depending on the current inventory. The wholesaler delivers exclusively to the central warehouse of the distributor.

The goal is to minimize inventory, while at the same time ensuring delivery.

Since the demand of the entire supply chain is not transparent, there is no forecast, but the warehouse is managed with expensive buffer capacities.

16.1.7   Green Beer Factory

The factory has no direct sales to end customers, but supplies the wholesaler exclusively. It provides the entire portfolio of beer products only.

Orders are accepted directly from the wholesaler and orders from the factory are sent to the production once per week depending on the current inventory.

The goal is to minimize inventory, while at the same time ensuring delivery.

Since the demand of the entire supply chain is not transparent, there is no forecast, but the warehouse is managed with expensive buffer capacities. The factory delivers exclusively to the central warehouse of the wholesaler. The production has unlimited capacity and can deliver demand on a weekly basis. Unexpected additional demand of the factory can be fulfilled within 7 days.

16.2   Green Beer IT: Infrastructure

16.2.1   Green Beer Retailer 1

There is no order management, no warehouse administration and no human resource management, only a PC-based financial system for processing invoices.

Orders are retrieved from the stores by telephone, whereby each sales person manages these data in a separate PC.

The orders are bundled and then sent directly to the distributor by email.

There is no transparency and there are no systems for demand planning and forecasting and no consistent centralized processes for the retail shops.

16.2.2   Green Beer Retailer 2

There is no order management, no warehouse administration and no human resource management, only a PC-based financial system for processing invoices.

Orders are retrieved from the markets by telephone, whereby each sales person manages these data in a separate PC.

The orders are bundled and then sent directly to the distributor by email.

There is no transparency and there are no systems for demand planning and forecasting and no consistent centralized processes for the markets.

16.2.3   Green Beer Retailer 3

There is no order management, no warehouse administration and no human resource management, only a PC-based financial system for processing invoices.

Orders are retrieved from the stores by telephone, whereby each sales person manages these data in a separate PC.

The orders are bundled and then sent directly to the distributor by email.

There is no transparency and there are no systems for demand planning and forecasting and no consistent centralized processes for the shops.

16.2.4   Green Beer Retailer 4

There is no order management, no warehouse administration and no human resource management, only a PC-based financial system for processing invoices.

Orders are retrieved from the stores by telephone, whereby each sales person manages these data in a separate PC.

The orders are bundled and then sent directly to the distributor by email.

There is no transparency and there are no systems for demand planning and forecasting and no consistent centralized processes for the stores.

16.2.5   Green Beer Distributor

There is a separate order management system, a small warehouse management system and a human resource management system, integrated into a PC-based financial system.

Orders are retrieved from the retailers by email. They are synchronized manually with the inventory and order to the wholesaler is entered using the factory's online portal.

There is no transparency and there are no systems for demand planning and forecasting and no consistent centralized processes order management and inventory management.

16.2.6   Green Beer Wholesaler

There is a separate order management system, a small warehouse management system and a human resource management system, integrated into a PC-based financial system.

Orders are retrieved from the distributor through the factory's online portal. They are synchronized manually with the inventory and orders to the factory is entered using the factory's online portal.

There is no transparency and there are no systems for demand planning and forecasting and no consistent centralized processes order management and inventory management.

16.2.7   Green Beer Factory

Production is largely automated. There is an ERP system for production planning, finance, human resource management (HR) and customer relationship management (CRM). CRM is used exclusively for order management.

Orders from the wholesaler are placed directly through the factory's online portal.

Supply chain management (SCM) was not considered necessary so far.

The ERP system is 8 years old and a major upgrade is planned, but is on hold due to the merger.

Chapter 17
Post-Merger Situation: Royal Beer

Abstract The chapter contains the necessary information about the market situation, distribution, customer structure and product portfolio as well as the IT infrastructure for each division directly after the merger.

17.1 Royal Beer Market: Distribution Structure—Customers

Figure 17.1 shows the Royal Beer product portfolios for each retail chain after the merger.

Previous year results (Figs. 17.2, 17.3, and 17.4).

17.1.1 Royal Beer Retailer 1

The retail chain 1 supplies its own stores and large retail chains with the entire portfolio mostly except kegs. In addition to beverages in their own shops, other goods, especially snacks are offered.

Orders from individual stores are collected and sent to the distributor once per week depending on demand and inventory.

The goal is to minimize inventory, while at the same time ensuring delivery. There is a central warehouse supporting all shops. The distributor delivers exclusively to the central warehouse.

There is no demand planning, but a high expectation to the distributor for immediate and complete delivery.

17.1.2 Royal Beer Retailer 2

Retail chain 2 supplies its own beverage markets with the entire portfolio except single bottles. Cases and kegs are delivered to private customers and restaurants only.

© Springer-Verlag GmbH Germany 2017
K.-D. Gronwald, *Integrated Business Information Systems*,
DOI 10.1007/978-3-662-53291-1_17

Fig. 17.1 Royal Beer product portfolio

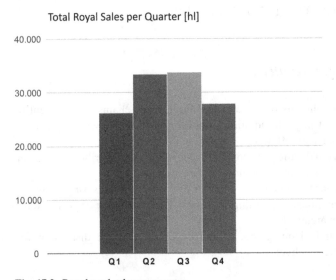

Fig. 17.2 Royal total sales per quarter

Orders from individual beverage markets are collected and sent to the distributor once per week depending on demand and inventory.

The goal is to minimize inventory, while at the same time ensuring delivery. There is a central warehouse supporting all shops. The distributor delivers exclusively to the central warehouse.

Total Sales per Retail Chain [hl]

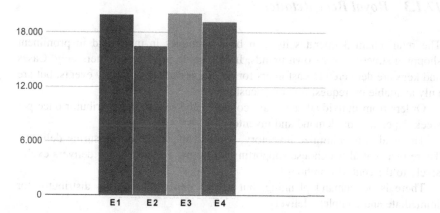

Fig. 17.3 Royal total sales per retail chain

Total EBITDA [%] per Retail Chain

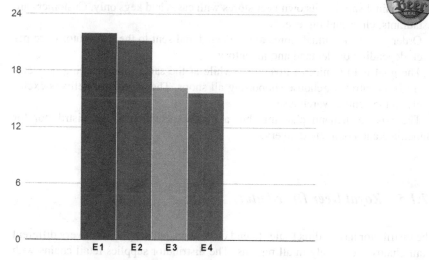

Fig. 17.4 Royal EBITDA per retail chain

There is no demand planning, but a high expectation to the distributor for immediate and complete delivery.

17.1.3   Royal Beer Retailer 3

The retail chain 3 operates its own beer boutiques in malls and in prominent shopping streets with its own brands. It delivers to private customers only. Cases and kegs are delivered to customers for special private or company events, but are only available on request for select customers.

Orders from individual stores are collected and sent to the distributor once per week depending on demand and inventory.

The goal is to minimize inventory, while at the same time ensuring delivery. There is a central warehouse supporting all shops. The distributor delivers exclusively to the central warehouse.

There is no demand planning, but a high expectation to the distributor for immediate and complete delivery.

17.1.4   Royal Beer Retailer 4

Retail chain 4 supplies its own beer stores with cases and kegs only. Customers are restaurants, clubs and big events.

Orders from individual stores are collected and sent to the distributor once per week depending on demand and inventory.

The goal is to minimize inventory, while at the same time ensuring delivery. There is a central warehouse supporting all shops. The distributor delivers exclusively to the central warehouse.

There is no demand planning, but a high expectation to the distributor for immediate and complete delivery.

17.1.5   Royal Beer Distributor

The distributor has no direct sales to end customers, but supplies four very different retail chains exclusively in all regions. The distributor supplies retail chains with soft drinks and bottled water additional to beer.

Orders are accepted directly from the retail chains and orders from the distributor are sent to the wholesaler once per week depending on the current inventory. The distributor delivers exclusively to the central warehouse of the retailers.

The goal is to minimize inventory, while at the same time ensuring delivery.

Since the demand of the entire supply chain is not transparent, there is no forecast, but the warehouse is managed with expensive buffer capacities.

17.1.6   Royal Beer Wholesaler

The wholesaler has no direct sales to end customers, but supplies the distributor exclusively. The wholesaler supplies distributors with soft drinks and bottled water additional to beer.

Orders are accepted directly from the distributor and orders from the wholesaler are sent to the factory once per week depending on the current inventory. The wholesaler delivers exclusively to the central warehouse of the distributor.

The goal is to minimize inventory, while at the same time ensuring delivery.

Since the demand of the entire supply chain is not transparent, there is no forecast, but the warehouse is managed with expensive buffer capacities.

17.1.7   Royal Beer Factory

The factory has no direct sales to end customers, but supplies the wholesaler exclusively. It provides the entire portfolio of beer products only.

Orders are accepted directly from the wholesaler and orders from the factory are sent to the production once per week depending on the current inventory.

The goal is to minimize inventory, while at the same time ensuring delivery.

Since the demand of the entire supply chain is not transparent, there is no forecast, but the warehouse is managed with expensive buffer capacities. The factory delivers exclusively to the central warehouse of the wholesaler. The production has unlimited capacity and can deliver demand on a weekly basis. Unexpected additional demand of the factory can be fulfilled within 7 days.

17.2   Royal Beer IT: Infrastructure

17.2.1   Royal Beer Retailer 1

There is no order management, no warehouse administration and no human resource management, only a PC-based financial system for processing invoices.

Orders are retrieved from the stores by telephone, whereby each sales person manages these data in a separate PC.

The orders are bundled and then sent directly to the distributor by email.

There is no transparency and there are no systems for demand planning and forecasting and no consistent centralized processes for the retail shops.

17.2.2   Royal Beer Retailer 2

There is no order management, no warehouse administration and no human resource management, only a PC-based financial system for processing invoices.

Orders are retrieved from the markets by telephone, whereby each sales person manages these data in a separate PC.

The orders are bundled and then sent directly to the distributor by email.

There is no transparency and there are no systems for demand planning and forecasting and no consistent centralized processes for the markets.

17.2.3   Royal Beer Retailer 3

There is no order management, no warehouse administration and no human resource management, only a PC-based financial system for processing invoices.

Orders are retrieved from the stores by telephone, whereby each sales person manages these data in a separate PC.

The orders are bundled and then sent directly to the distributor by email.

There is no transparency and there are no systems for demand planning and forecasting and no consistent centralized processes for the shops.

17.2.4   Royal Beer Retailer 4

There is no order management, no warehouse administration and no human resource management, only a PC-based financial system for processing invoices.

Orders are retrieved from the stores by telephone, whereby each sales person manages these data in a separate PC.

The orders are bundled and then sent directly to the distributor by email.

There is no transparency and there are no systems for demand planning and forecasting and no consistent centralized processes for the stores.

17.2.5 Royal Beer Distributor

There is a separate order management system, a small warehouse management system and a human resource management system, integrated into a PC-based financial system.

Orders are retrieved from the retailers by email. They are synchronized manually with the inventory and order to the wholesaler is entered using the factory's online portal.

There is no transparency and there are no systems for demand planning and forecasting and no consistent centralized processes order management and inventory management.

17.2.6 Royal Beer Wholesaler

There is a separate order management system, a small warehouse management system and a human resource management system, integrated into a PC-based financial system.

Orders are retrieved from the distributor through the factory's online portal. They are synchronized manually with the inventory and orders to the factory is entered using the factory's online portal.

There is no transparency and there are no systems for demand planning and forecasting and no consistent centralized processes order management and inventory management.

17.2.7 Royal Beer Factory

Production is largely automated. There is an ERP system for production planning, finance, human resource management (HR) and customer relationship management (CRM). CRM is used exclusively for order management.

Orders from the wholesaler are placed directly through the factory's online portal.

Supply chain management (SCM) was not considered necessary so far.

The ERP system is 8 years old and a major upgrade is planned, but is on hold due to the merger.

Chapter 18
Post-Merger Situation: Wild Horse Beer

Abstract The chapter contains the necessary information about the market situation, distribution, customer structure and product portfolio as well as the IT infrastructure for each division directly after the merger.

18.1 Wild Horse Beer Market: Distribution Structure— Customers

Figure 18.1 shows the Wild Horse Beer product portfolios for each retail chain after the merger.

Previous year results (Figs. 18.2, 18.3, and 18.4).

18.1.1 Wild Horse Beer Retailer 1

The retail chain 1 supplies its own stores and large retail chains with the entire portfolio mostly except kegs. In addition to beverages in their own shops, other goods, especially snacks are offered.

Orders from individual stores are collected and sent to the distributor once per week depending on demand and inventory.

The goal is to minimize inventory, while at the same time ensuring delivery. There is a central warehouse supporting all shops. The distributor delivers exclusively to the central warehouse.

There is no demand planning, but a high expectation to the distributor for immediate and complete delivery.

18.1.2 Wild Horse Beer Retailer 2

Retail chain 2 supplies its own beverage markets with the entire portfolio except single bottles. Cases and kegs are delivered to private customers and restaurants only.

© Springer-Verlag GmbH Germany 2017
K.-D. Gronwald, *Integrated Business Information Systems*,
DOI 10.1007/978-3-662-53291-1_18

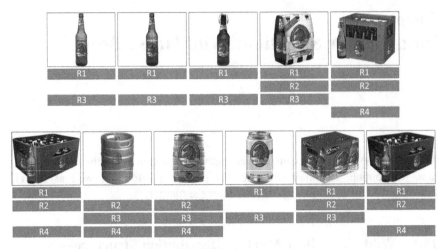

Fig. 18.1 Wild Horse Beer product portfolio

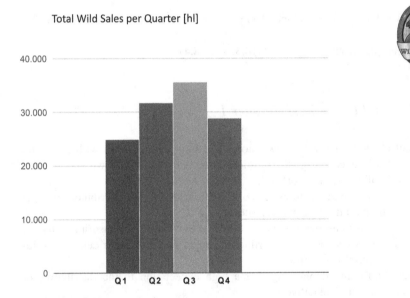

Fig. 18.2 Wild Horse total sales per quarter

Orders from individual beverage markets are collected and sent to the distributor once per week depending on demand and inventory.

The goal is to minimize inventory, while at the same time ensuring delivery. There is a central warehouse supporting all shops. The distributor delivers exclusively to the central warehouse.

There is no demand planning, but a high expectation to the distributor for immediate and complete delivery.

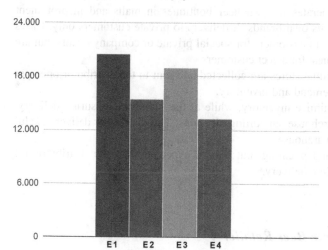

Fig. 18.3 Wild Horse total sales per retail chain

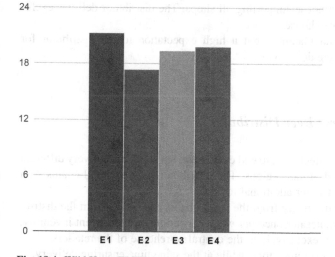

Fig. 18.4 Wild Horse EBITDA per retail chain

18.1.3   Wild Horse Beer Retailer 3

The retail chain 3 operates its own beer boutiques in malls and in prominent shopping streets with its own brands. It delivers to private customers only. Cases and kegs are delivered to customers for special private or company events, but are only available on request for select customers.

Orders from individual stores are collected and sent to the distributor once per week depending on demand and inventory.

The goal is to minimize inventory, while at the same time ensuring delivery. There is a central warehouse supporting all shops. The distributor delivers exclusively to the central warehouse.

There is no demand planning, but a high expectation to the distributor for immediate and complete delivery.

18.1.4   Wild Horse Beer Retailer 4

Retail chain 4 supplies its own beer stores with cases and kegs only. Customers are restaurants, clubs and big events.

Orders from individual stores are collected and sent to the distributor once per week depending on demand and inventory.

The goal is to minimize inventory, while at the same time ensuring delivery. There is a central warehouse supporting all shops. The distributor delivers exclusively to the central warehouse.

There is no demand planning, but a high expectation to the distributor for immediate and complete delivery.

18.1.5   Wild Horse Beer Distributor

The distributor has no direct sales to end customers, but supplies four very different retail chains exclusively in all regions. The distributor supplies retail chains with soft drinks and bottled water additional to beer.

Orders are accepted directly from the retail chains and orders from the distributor are sent to the wholesaler once per week depending on the current inventory. The distributor delivers exclusively to the central warehouse of the retailers.

The goal is to minimize inventory, while at the same time ensuring delivery.

Since the demand of the entire supply chain is not transparent, there is no forecast, but the warehouse is managed with expensive buffer capacities.

18.1.6   Wild Horse Beer Wholesaler

The wholesaler has no direct sales to end customers, but supplies the distributor exclusively. The wholesaler supplies distributors with soft drinks and bottled water additional to beer.

Orders are accepted directly from the distributor and orders from the wholesaler are sent to the factory once per week depending on the current inventory. The wholesaler delivers exclusively to the central warehouse of the distributor.

The goal is to minimize inventory, while at the same time ensuring delivery.

Since the demand of the entire supply chain is not transparent, there is no forecast, but the warehouse is managed with expensive buffer capacities.

18.1.7   Wild Horse Beer Factory

The factory has no direct sales to end customers, but supplies the wholesaler exclusively. It provides the entire portfolio of beer products only.

Orders are accepted directly from the wholesaler and orders from the factory are sent to the production once per week depending on the current inventory.

The goal is to minimize inventory, while at the same time ensuring delivery.

Since the demand of the entire supply chain is not transparent, there is no forecast, but the warehouse is managed with expensive buffer capacities. The factory delivers exclusively to the central warehouse of the wholesaler. The production has unlimited capacity and can deliver demand on a weekly basis. Unexpected additional demand of the factory can be fulfilled within 7 days.

18.2   Wild Horse Beer IT: Infrastructure

18.2.1   Wild Horse Beer Retailer 1

There is no order management, no warehouse administration and no human resource management, only a PC-based financial system for processing invoices.

Orders are retrieved from the stores by telephone, whereby each sales person manages these data in a separate PC.

The orders are bundled and then sent directly to the distributor by email.

There is no transparency and there are no systems for demand planning and forecasting and no consistent centralized processes for the retail shops.

18.2.2   Wild Horse Beer Retailer 2

There is no order management, no warehouse administration and no human resource management, only a PC-based financial system for processing invoices.

Orders are retrieved from the markets by telephone, whereby each sales person manages these data in a separate PC.

The orders are bundled and then sent directly to the distributor by email.

There is no transparency and there are no systems for demand planning and forecasting and no consistent centralized processes for the markets.

18.2.3   Wild Horse Beer Retailer 3

There is no order management, no warehouse administration and no human resource management, only a PC-based financial system for processing invoices.

Orders are retrieved from the stores by telephone, whereby each sales person manages these data in a separate PC.

The orders are bundled and then sent directly to the distributor by email.

There is no transparency and there are no systems for demand planning and forecasting and no consistent centralized processes for the shops.

18.2.4   Wild Horse Beer Retailer 4

There is no order management, no warehouse administration and no human resource management, only a PC-based financial system for processing invoices.

Orders are retrieved from the stores by telephone, whereby each sales person manages these data in a separate PC.

The orders are bundled and then sent directly to the distributor by email.

There is no transparency and there are no systems for demand planning and forecasting and no consistent centralized processes for the stores.

18.2.5   Wild Horse Beer Distributor

There is a separate order management system, a small warehouse management system and a human resource management system, integrated into a PC-based financial system.

Orders are retrieved from the retailers by email. They are synchronized manually with the inventory and order to the wholesaler is entered using the factory's online portal.

There is no transparency and there are no systems for demand planning and forecasting and no consistent centralized processes order management and inventory management.

18.2.6   Wild Horse Beer Wholesaler

There is a separate order management system, a small warehouse management system and a human resource management system, integrated into a PC-based financial system.

Orders are retrieved from the distributor through the factory's online portal. They are synchronized manually with the inventory and orders to the factory is entered using the factory's online portal.

There is no transparency and there are no systems for demand planning and forecasting and no consistent centralized processes order management and inventory management.

18.2.7   Wild Horse Beer Factory

Production is largely automated. There is an ERP system for production planning, finance, human resource management (HR) and customer relationship management (CRM). CRM is used exclusively for order management.

Orders from the wholesaler are placed directly through the factory's online portal.

Supply chain management (SCM) was not considered necessary so far.

The ERP system is 8 years old and a major upgrade is planned, but is on hold due to the merger.

Chapter 19
Share of Wallet

Abstract This chapter contains detailed information about the initial share of wallet for all four beer groups and the four key accounts KDISCOUNT, KDISUPER, KDIvalue, and KDIFRESH.

19.1 Share of Wallet Alpha Beer

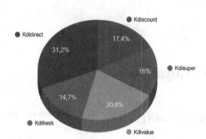

ALPHA Customers [%]   Alpha Share of Wallet

Customer	Revenue/year	Margin	Discount	Lifetime	NPV
KDISCOUNT	$7'325'423	16.97%	8.0%	8 years	7'145'891
KDISUPER	$6'761'929	20.07%	4.0%	5 years	6'041'552
KDIvalue	$8'707'578	18.51%	6.0%	3 years	4'308'869
KDIFRESH	$6'219'699	18.53%	6.0%	7 years	6'434'634

19.2   Share of Wallet Green Beer

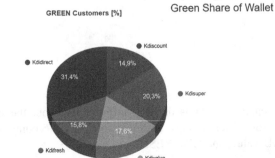

GREEN Customers [%]   Green Share of Wallet

Customer	Revenue/year	Margin	Discount	Lifetime	NPV
KDISCOUNT	$6'515'209	19.28%	5.0%	3 years	3'420'981
KDISUPER	$8'867'923	17.03%	8.0%	4 years	5'003'148
KDIvalue	$7'672'115	17.74%	7.0%	6 years	6'488'790
KDIFRESH	$6'883'995	18.57%	6.0%	5 years	5'385'002

19.3   Share of Wallet Royal Beer

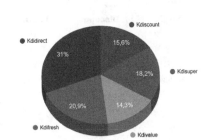

ROYAL Customers [%]   Royal Share of Wallet

Customer	Revenue/year	Margin	Discount	Lifetime	NPV
KDISCOUNT	$7'069'050	18.51%	6.0%	4 years	4'534'638
KDISUPER	$8'264'552	21.59%	2.0%	7 years	11'549'728
KDIvalue	$6'504'508	16.97%	8.0%	5 years	4'408'511
KDIFRESH	$9'456'554	20.06%	4.0%	6 years	9'942'116

19.4   Share of Wallet Wild Horse Beer

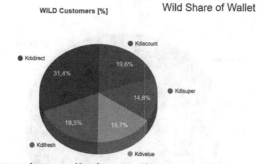

Customer	Revenue/year	Margin	Discount	Lifetime	NPV
KDISCOUNT	$8'166'470	13.90%	12.0%	4 years	3'447'814
KDISUPER	$6'163'374	18.45%	6.0%	3 years	3'039'720
KDIvalue	$6'512'823	17.74%	7.0%	2 years	2'089'380
KDIFRESH	$7'686'089	21.53%	2.0%	5 years	7'799'211

19.5   Share of Wallet KDISUPER

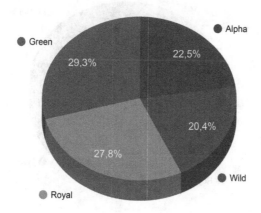

19.6   Share of Wallet **KDIFRESH**

KDIFRESH Wallet [%]

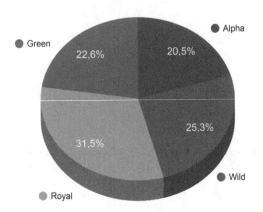

19.7   Share of Wallet **KDISCOUNT**

KDISCOUNT Wallet [%]

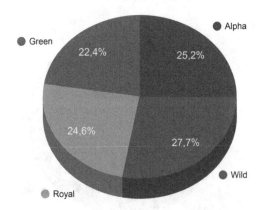

19.8   Share of Wallet KDIvalue

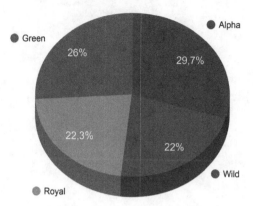

Index

© Springer-Verlag GmbH Germany 2017
K.-D. Gronwald, *Integrated Business Information Systems*,
DOI 10.1007/978-3-662-53291-1

Printed in the United States
By Bookmasters